Adaptations

SUSAN SCHAFER

Sharpe Focus
An imprint of M.E. Sharpe, Inc.
80 Business Park Drive
Armonk, NY 10504
www.sharpe-focus.com

Library of Congress Cataloging-in-Publication Data

Schafer, Susan.
Adaptations / Susan Schafer.
p. cm. -- (Genetics: the science of life)
Includes bibliographical references and index.
ISBN 978-0-7656-8137-9 (hardcover : alk. paper)
1. Adaptation (Biology)--Juvenile literature. I. Title.
QH546.S33 2009
578.4--dc22

2008008106

Editor: Peter Mavrikis
Production Manager: Henrietta Toth
Editorial Assistant and Photo Research: Alison Morretta
Program Coordinator: Cathy Prisco
Design: Patrice Sheridan
Line Art: FoxBytes
Printed in Malaysia

9 8 7 6 5 4 3 2 1

PICTURE CREDITS: Cover (left): National Geographic/Getty Images; cover (right): Science Faction/Getty Images; pages 4, 49 (top right): Gallo Images/Getty Images; pages 6, 11, 25: FoxBytes; pages 12, 45 (bottom), 66: Photonica/Getty Images; pages 13, 18 (bottom left), 22, 28, 29, 43, 51 (left), 59 (right), 63, 65, 77, 79 (right), 81, 84 (top left): National Geographic/Getty Images; page 14: WireImage/Getty Images; page 18 (top left): All Canada Photos/Getty Images; pages 18 (right), 59 (left), 70, 76 (top far right): Photographer's Choice/Getty Images; pages 19, 76 (bottom), 84 (bottom left): Discovery Channel Images/Getty Images; page 20: Time & Life Pictures/Getty Images; pages 27, 49 (bottom right), 72: AFP/Getty Images; pages 31, 46, 47 (left), 51 (right), 54, 58 (left), 61, 76 (top far left): Visuals Unlimited/Getty Images; pages 32, 36, 47 (middle), 50: Science Faction/Getty Images; pages 38 (bottom left), 44, 48, 76 (top middle), 84 (right), 86: Dorling Kindersley/Getty Images; pages 38 (top left), 49 (left), 56, 60, 74: Stone/Getty Images; pages 38 (right), 40, 41, 47 (right): Photolibrary/Getty Images; pages 39, 79 (left): Taxi/Getty Images; pages 45 (top), 58 (right), 68: Robert Harding World Imagery/Getty Images; pages 52, 76 (top middle right): © iStock.com; page 53: Altrendo/Getty Images; page 62: ScienceFoto/Getty Images; page 76 (top middle left): Getty Images; page 87: Asia Images/Getty Images; back cover: Photographer's Choice/Getty Images.

Contents

Elephants, zebras, and antelope exhibit the genetic variation that is found in all living organisms.

CHAPTER 1

Adaptations

You are hiking in the desert. The morning air is fresh and the sand is blowing softly against the rocks. Soon, the air heats up and the sun blazes down on you. You are thirsty, but you did not bring water. You turn back, realizing how far you have come. It will be a long, hot walk, but your body has adaptations that will help.

Your face flushes red with blood, which moves close to the surface of your skin to release heat. Sweat glands release cooling water onto your skin, which evaporates and carries the heat away from your body. If it gets too hot, chemical changes in your body signal your brain that your temperature is too high. Your brain then signals your muscles to move into shade.

FIT THE ENVIRONMENT

An adaptation is any characteristic that ensures an individual is fit for living under the conditions of its environment. Without **adaptations**, organisms would not survive. The same adaptations that help protect humans in the desert keep their body temperature normal wherever they

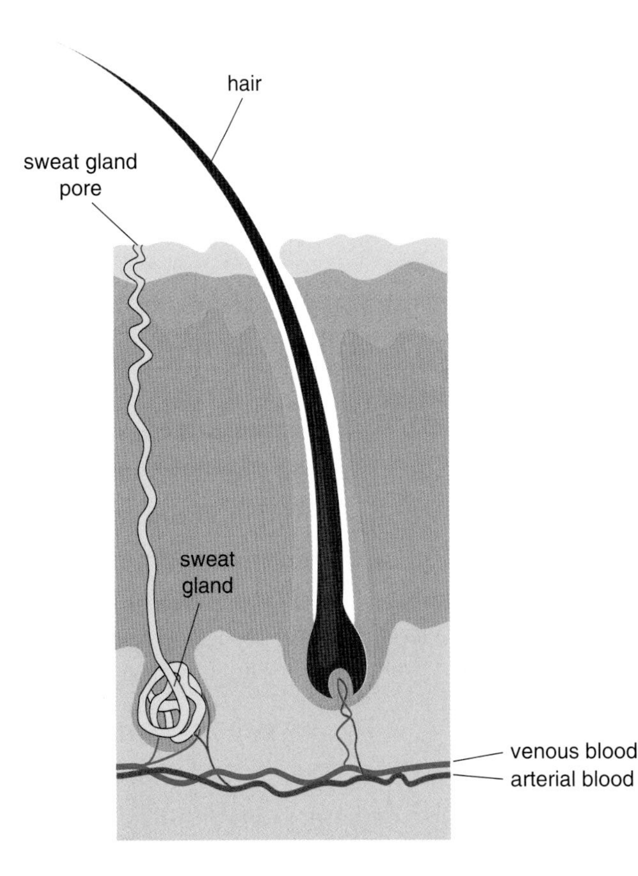

Sweat glands secrete sweat, an adaptation that cools the body when it becomes overheated. A person has more than 2.5 million sweat glands distributed over the entire surface of the skin.

are. Maintaining a normal body temperature is important because if body cells get too hot, chemical reactions do not work properly and the cells die. If body cells get too cold, chemical reactions slow down and the cells stop working altogether.

Scientists estimate that between 5 million and 30 million kinds, or **species**, of organisms live on Earth. A species is a group of living things that can mate with one another and produce young that can also grow up to mate. Only about 2 million of the estimated species have been discovered and described by scientists. But discovered or not, every organism has adaptations that help it survive.

The variation that exists in all of the living things on Earth creates **biodiversity**. Biodiversity includes **species diversity**, which describes the differences that are found within a species or among many different species. These differences can be in how each of them looks, how they function or act, or how they react to other species in their environment. Biodiversity also includes the variation that occurs in the genes of a species (**genetic diversity**) or in a particular habitat on Earth (**ecosystem diversity**).

Scientists believe that greater biodiversity in a region, or the number of different plants and animals that live there, indicates a healthy environment. Different species depend on each other for food, shelter, and so on,

Classification is the arrangement of things into groups based on their similarities. When things are classified together, it indicates some kind of relationship between all the members of the group. For example, all reptiles are related because they all have scales. Scientists group organisms into seven levels of classification, based on their shared characteristics. The largest group is called the kingdom. The kingdom is divided into smaller groups called phyla *(FY-luh)* (*singular*, phylum). Each group continues to be divided into smaller and smaller groups: class, order, family, genus, and finally species. The species level contains only one unique kind of organism. The more classification levels that two organisms share, the more closely related they are.

KINGDOM
Animalia — organisms with eukaryotic cells containing a nucleus and no cell wall

PHYLUM
Chordata — animals with a notochord for support of the body

CLASS
Mammalia — warm-blooded chordates with a backbone, hair, and mammary glands for producing milk

ORDER
Primates — mammals with five fingers, fingernails, an opposable thumb, a distinct dental pattern, and special bones in the eye sockets

FAMILY
Hominidae — primates without tails

GENUS
Homo — hominids with little body hair

SPECIES
sapiens — species of Homo with a highly developed brain and language

which is one reason it is important to preserve all of Earth's species.

Biodiversity exists because of genes, which give each species its unique adaptations. When organisms have most of the same genes in common and therefore the same characteristics, scientists recognize them as the same species. In naming species, scientists organize, or **classify**, groups of animals using Latin terms to describe them. For example, the Asian elephant belongs to a species called *Elephas maximus*. The first word, *Elephas*, is actually the **genus** *(JEE-nuhs)* (*plural*, genera), which is the larger scientific grouping of elephants. The second word, *maximus*, indicates the species name.

People belong to the species *Homo sapiens (HOE-moe SAY-pee-ehnz)*. About 99 percent of each person's genes are exactly the same as all other

DOWNLOAD

- Classification is the arrangement of living or nonliving things into groups based on their shared characteristics.
- An adaptation is a characteristic that helps an organism survive in its environment.
- Natural selection is the process by which organisms with favorable genes survive long enough to reproduce and pass on those genes.
- A species is the smallest division of living things.
- Within a species that reproduces sexually, individuals can mate with others of their kind and produce young that can also grow up to mate.
- A vestigial structure, such as the human appendix, is a remnant of a once-useful adaptation.
- Artificial selection is the process by which humans choose desired traits in a species and allow only individuals carrying those traits to reproduce.

humans. The remaining 1 percent makes each person unique. The system of using two names to describe an organism is called **binomial nomenclature** *(by-NOH-mee-uhl NOH-men-klay-chur)*, which was started in the 1750s by a Swedish scientist named Carolus Linnaeus.

DO OR DIE

An organism is adapted to its environment when it can find food, water, a place to live, and a mate. It can deal with heat, cold, wind, and rain and defend itself from predators. These are some of the characteristics that define a thing as living, as opposed to a rock or dirt. Living things use energy from food, grow, respond to the environment, and reproduce.

Adaptations are inherited through a process called **natural selection**. In other words, what happens in nature or in the environment selects or determines which genes or traits will be passed on to each new generation. When individuals are born, each has a unique set of genes. Some individuals have genes that make it more likely they will survive.

TOOL BAR

Natural selection is the tool of adaptation. It is what makes it happen. Think of it as a cycle. First, a species produces more offspring than can survive. Some will have favorable genes and some will not. All of the offspring struggle to survive and, because of their genes, some will make it and some will not. Finally, any survivors that reproduce will pass on their genes. Then it is back to step one. The cycle repeats for generations and slowly the species adapts to its environment.

All young that are born will struggle to survive in their environment. Only those that find food, water, shelter, and other things necessary for survival will live long enough to reproduce and pass on their genes. The genes of those that do not survive die with the individual. Slowly over time, generation after generation, more and more individuals inherit the successful traits. In this way, the species adapts to its environment.

IF YOU DO NOT NEED IT, LOSE IT

Not every characteristic in a species is an adaptation. Most species have some genes that produce traits that may no longer be useful. For example, humans still have a tail. It is short and hidden, but it is still there. It is referred to as the coccyx, or tailbone. Humans share this characteristic with tailless apes.

The human tailbone is **vestigial** *(veh-STIJ-ee-uhl)*, which means that the tailbone has remained in some form even though it no longer functions as it did for early, less humanlike ancestors. Those ancestors needed their tails in order to balance. The tail's secondary function (the use that was second to balancing) was to allow mobility. This secondary function is now the tailbone's only function. Muscles in the behind attach to the tailbone, allowing the legs to move properly for sitting, walking, or running.

Besides the tail, people have vestigial teeth called wisdom teeth, a vestigial appendix, and vestigial structures in the skin that cause goose bumps. Wisdom teeth were once big, flat molars that were used to grind tough plant material before human ancestors became carnivores. The genes for wisdom teeth may some day completely disappear from the human population. In the meantime, dentists may continue to pull them out.

The appendix was once used to digest tough plant fibers. If the appendix becomes infected, it can be removed just as wisdom teeth are. Because it can be removed without causing any harm, doctors once believed the appendix had no function at all. Recently, however, scientists discovered that the appendix functions secondarily as a storeroom for beneficial gut bacteria. If an illness destroys the normal bacteria that live in the intestines and help digestion, the appendix can send out a new supply.

Goose bumps are caused by a tiny muscle that is attached to each of the hairs in the skin. When the muscle contracts, it pulls on the hair and it stands up straight, pushing the skin into a little bulge—a goose bump. By raising the hairs in this way, hairy human ancestors kept warmer in the cold and looked bigger to scare off enemies. Cats do this when they raise the hairs on their backs.

Many other species have vestigial structures that no longer function, have a reduced function, or function differently. Whales have vestigial hind legs that can be seen in X-rays. Tiny bones on either side of a whale's pelvis were once working legs. But these hind legs created drag and kept them from moving swiftly in the water. Individuals without rear legs were more likely to survive, as they were able to catch food and escape predators more quickly. These survivors then passed on the "no–rear-legs" genes to modern whales.

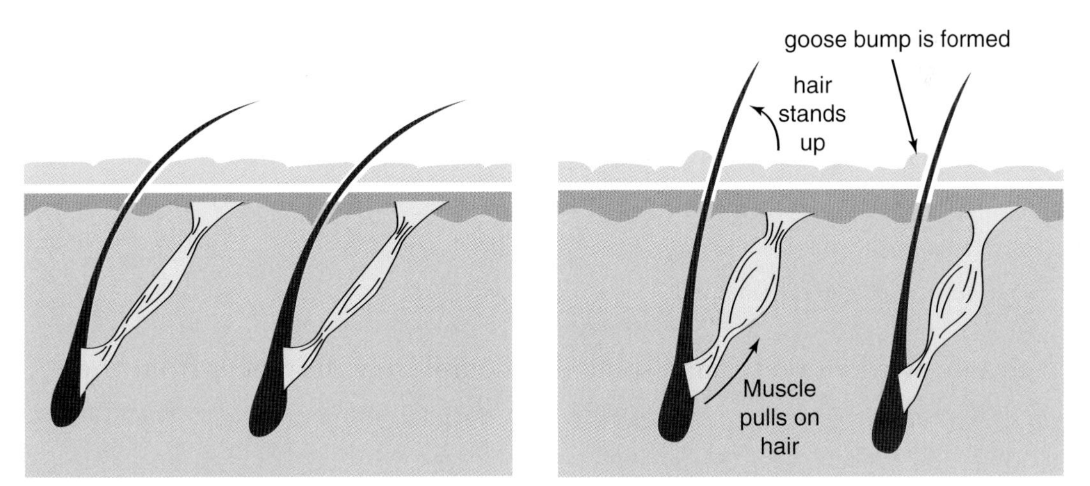

The contraction of a tiny muscle attached to the base of a hair causes the hair to stand up, pushing up a bump of skin called a goose bump or goose flesh.

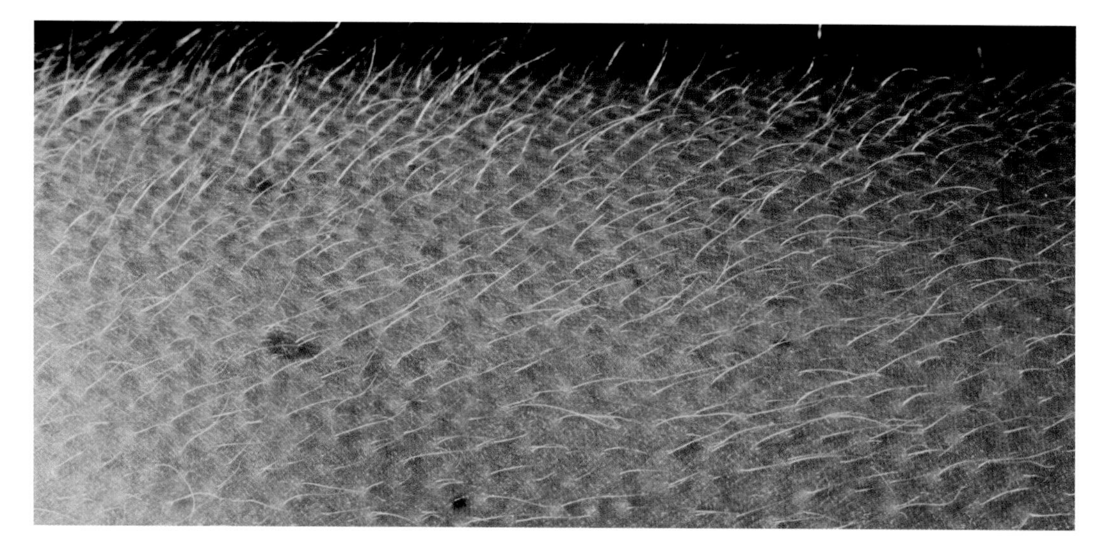

Goose bumps are formed when the hairs in human skin stands up in response to cold or fear. In animals with fur, the lifted hairs make the animals fluffier, which traps warm air close to the body.

Animals that live permanently in caves, such as certain fish and salamanders, have vestigial eyes, which may be nothing more than small black dots. The eyes may be able to detect light or dark, but they cannot form images. Sight is not necessary for these creatures because they live in total darkness. Their other senses became adapted over time to help them to survive in this environment. Those with a better sense of touch or smell lived and then passed on those genes. Those with a better sense of sight were not as successful at finding food or escaping predators, so they died out.

FLUKES OF NATURE

Sometimes adaptations start as one thing and become another. Feathers began as an adaptation for insulation, to keep birds warm. Then they evolved to allow birds to fly, but not all birds use them for this. Penguins use their wings as flippers to swim underwater. Male ostriches use their tiny, vestigial wings to swish their feathers and twist their bodies in a bizarre dance that will hopefully attract mates.

The young of a South American bird called the hoatzin (*waht-SIN*) use their wings to climb. Each wing has two claws. When threatened by a predator, baby hoatzins, which cannot yet fly, jump from the nest and fall to the ground. When it is safe, they use their wings and claws to climb back to the nest. Once the hoatzins mature and are able to fly, the claws disappear.

Some traits are passed on accidentally. An individual might get lucky for one reason or another and survive to pass on less desirable traits to its offspring. Perhaps an organism's ancestor was a slow runner but happened to turn left when turning right would have meant getting eaten by a predator. If that surviving individual was then able to mate, it would pass on "slow" genes. Scientists call these accidents **genetic drift**. A gene might become more or less frequent in a population simply by chance and not because it is

The chicks of the hoatzin have claws on the tips of their wings that allow them to climb and move about until their flight feathers come in.

POP-UP

Artificial selection is the process by which humans choose the males and females of a species that will be allowed to reproduce. In that way, people, instead of nature, choose which traits will be passed on to the offspring. People choose traits depending on the function or job the species is to perform for them. Over thousands of years, humans have artificially changed everything from crops and garden plants to livestock and pets. Some species—such as the domestic horse, *Equus caballus*—have been bred to serve humans as tools of war, workers, companions, and entertainers.

An ancestor of the horse, *Mesohippus,* ran wild around 35 million years ago. It stood about two feet (about half a meter) tall and had three toes on each foot. Over time, the three-toed horse evolved into a large, single-toed horse like the ones we know today. The single toe of the modern horse allows it to turn quickly and escape from predators.

For more than 6,000 years, humans have been domesticating horses, changing their size, build, color, markings, movement, and disposition. Horses have been bred to pull heavy wagons, plow, drive, perform, and carry riders for hunting, racing, jumping, and pleasure. They range in size from the tiniest miniature horse to the largest draft horse, and range in personality from agile and spirited to calm and patient. Today, official registries recognize over 300 breeds. Using artificial selection, humans have changed organisms like the horse far faster than they would have been changed by nature alone.

Through artificial selection, humans have developed the world's tallest horse—a Shire horse standing 6'9" at the withers (the top of the bump above the shoulders, at the bottom of the mane)—and the world's smallest horse, the miniature horse—which must stand less that 2'10" at the withers.

useful or adaptive. Genes that might be adaptive could also disappear by the process of genetic drift, leaving a population forever by chance.

Humans have produced their own kind of genetic drift, creating species that would not survive in the wild. By manipulating the genes of **domesticated** plants and animals, they make products that are more desirable, such as sweeter corn or rosebushes without thorns. Instead of nature selecting the genes, people do. For example, people wanted smaller dogs, so they bred Chihuahuas. For those who prefer bigger dogs, they bred St. Bernards. This is referred to as **artificial selection**, or selective breeding. People choose genes by allowing only certain individuals to breed. Whether or not the genes are adaptive is not a consideration, although sometimes a species is called "improved." An improved crop plant, for example, might be more resistant to insects or drought, which would help farmers.

Unfortunately, sometimes genes chosen for one reason can carry harmful genes on the same chromosomes. Both the desired trait and a harmful trait are passed on together. For example, German shepherd dogs often have severe hip problems. Blue merle Australian shepherds with too much white fur have a tendency to be blind and deaf.

NOTHING IS EVER PERFECT

A species is never totally adapted to its environment because the environment is always changing. What might be an adaptive trait now might not be in the future. In addition, mutations can change the genetic makeup of a population. New characteristics might arise that have never occurred before. The ability of genes to mutate or change is what makes adaptation possible. Gene mutations give a population the variation needed for natural selection to proceed because individuals carrying any changes could be more or less likely to survive.

Genetic diversity or gene variation might be good for the survival of a whole species, but it is not necessarily beneficial to individuals. If every individual in a population had identical genes, a disease or change in the environment could affect all of them. If all of the individuals died, the species would become extinct. With diversity, some individuals will die, but

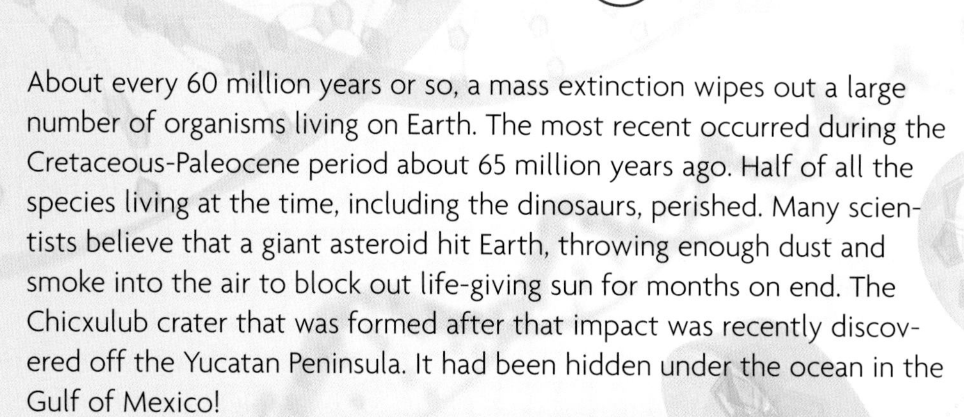

About every 60 million years or so, a mass extinction wipes out a large number of organisms living on Earth. The most recent occurred during the Cretaceous-Paleocene period about 65 million years ago. Half of all the species living at the time, including the dinosaurs, perished. Many scientists believe that a giant asteroid hit Earth, throwing enough dust and smoke into the air to block out life-giving sun for months on end. The Chicxulub crater that was formed after that impact was recently discovered off the Yucatan Peninsula. It had been hidden under the ocean in the Gulf of Mexico!

Another mass extinction occurred during the Triassic-Jurassic period around 200 million years ago. Massive lava flows spewed out of the land, opening the Atlantic Ocean and pushing apart the continents. Scientists believe the eruptions released toxic and heat-trapping gases that created runaway global warming. More than 50 percent of all ocean-dwelling genera died. Today, rocks from the eruption have been found in the eastern United States, eastern Brazil, Spain, and North Africa.

The worst of all extinctions occurred in the Permian-Triassic period about 250 million years ago. Ninety-five percent of all living species perished. Two additional mass extinctions have been recorded during Earth's history, one about 364 million years ago and another around 439 million years ago. The latter was first triggered when glaciers formed, causing a significant drop in sea level. Then, when the glaciers melted, sea levels rose dramatically.

Overall, more than 90 percent of all organisms that have ever lived on Earth are extinct. And while mass extinctions are extreme, they actually open up the environment for new life forms to evolve. The dinosaurs appeared after the Triassic-Jurassic extinctions. When the dinosaurs disappeared during the Cretaceous-Paleocene extinctions, the land was open for the diversification of mammals.

Another mass extinction is about due on Earth, but this time the cause may not be an asteroid impact or volcanic eruptions. Many worry that global warming caused by careless human activity will trigger the next annihilation.

those with genes that resist disease or a change to the environment would survive. Slowly over time, the population would increase again with more individuals carrying the adaptive or newly mutated genes. Individuals may be lost, but the species lives on.

An unfavorable gene, however, does not necessarily die out of a population. For example, most humans have a beneficial gene that allows them to taste a bitter chemical called phenylthiocarbamide *(fee-nuhl-thigh-oh-CAR-buh-myde)*, or PTC. PTC is a chemical found in many poisonous plants. If a boy carrying the "taster" gene were to taste PTC in a plant, he would spit it out.

But not all humans carry the gene that allows them to taste PTC in poisonous plants. They are called "nontasters." The gene they carry does not trigger the taste buds to respond to PTC, so nontasters are at an adaptive disadvantage. So how did their ancestors survive to pass on the gene that puts them at risk? It may have happened by genetic drift. It is possible that their ancestors simply never came across poisonous plants that carried PTC, but more likely it has to do with the fact that humans live in groups or family units. If those who were able to taste PTC made faces and spit out some juicy-looking fruit, any nontasters would probably not even try it. Or if they did, the tasters may have stopped them from eating it. In this way, the nontasters would survive, and the gene would stay in the population.

LOOK AND PLAY THE PART

Adaptations can be classified, or grouped, into three basic types: structural, physiological *(fizz-ee-uh-LAHJ-ih-kuhl)*, and behavioral. A structure is something that is built or put together in a certain way. Buildings are sometimes called structures, and as buildings are made of bricks, most body structures are made of **cells**. Horns for defense, claws for grabbing food, feathers for flying, skin for protection, toes for walking or climbing, teeth for chewing, and a large heart for running are all structural adaptations.

Physiology is the way cells, organs, and other body parts function or work chemically. When a person exercises, blood flow increases, which

Top left: The horns that bighorn sheep use to ram heads and establish dominance are structural adaptations. *Bottom left:* Sweating is a physiological adaptation that helps an organism maintain its normal body temperature. *Right:* The roar of a lion is a behavioral adaptation used to communicate with other lions, although the voice box in the lion's throat that allows it to roar is a structural adaptation.

brings more oxygen to the muscles. This blood-flow increase in response to exercise is a physiological adaptation. Any organism that lacks this adaptation would tire easily and might not survive. In the human body, maintaining temperature, sweating, making digestive juices in the stomach, and producing urine in the kidneys are all examples of physiological adaptations.

Behavioral adaptations affect the way a species acts. For example, a bird might dance to attract a mate, or a lizard might bob its head to threaten intruders in its territory. A person might smile at someone she thinks is attractive. Anything that increases the likelihood that an individual will succeed in life long enough to reproduce would be considered an adaptation.

Different species of lizards have different head-bobbing patterns that are determined by their genes. This helps the lizards identify members of their own species. Each type of lizard responds only to head bobs that are the same as their own. It is like a language. If you speak only English, you only understand other English speakers.

Like waving a flag, a male anole uses a colorful throat flap called a dewlap to communicate with other anoles. The dewlap, which is normally hidden, is erected using a special bone in the throat.

These behaviors are **innate**, or present at birth. They are also known as instincts. A rabbit has the instinct to run from a predator. If it had to take time to learn this behavior, it would not live long. Other behaviors, however, are learned from family or group members. A young lion must learn how to hunt by watching the lionesses. Some scientists, however, would argue that the ability to learn would have to be inherited in the genes and therefore would still be an adaptation.

BLOG

Linnaeus was proud of the work he was doing in the classification of living things and adopted the motto:

> God created, Linnaeus organized.
>
> —Linnaeus, 1707–1778

Carolus Linnaeus developed the system for classifying plants and animals that is still used today.

CAROLUS LINNAEUS

Carolus Linnaeus was a Swedish physician, botanist, and zoologist. He is often called the Father of Taxonomy. Taxonomy is the science of grouping or classifying things. The word comes from the Greek word *taxis*, which means "arrangement." In 1735, Linnaeus published his first edition of the classification of living things in Latin. He called it the *Systema Naturae per regna tria naturae: secumdum classes, ordines, genera, species cum characteribus, differentiis, synonymis, locis* (System of Nature through the three kingdoms of nature: according to classes, orders, genera, and species with characters, differences, synonyms, places). Some of the Latin words in the title may look familiar because many of the scientific terms used today are based on the Latin language.

Over the years, Linnaeus revised his work to include more and more plant and animal species, some that he collected during his own travels. He grouped species by the characteristics they had in common, placing them into levels of classification (kingdom, class, order, genus, species). Later, scientists determined additional levels of similarity, such as phylum below the level of kingdom and family below the level of order.

Before Linnaeus introduced his system, species names were long and cumbersome. Linnaeus simplified things by assigning each species a binomial, or two-part, Latin name. The first part of the name indicates the genus; the second name indicates the species. The binomial name applies only for one specific group of unique organisms. Much of the Linnaean system, such as the levels of classification and assigning binomial names, is still used by scientists.

A change in a gene, such as one that might alter the color of a plant's flower, might give one plant a survival advantage that other plants in the same environment might not have.

CHAPTER 2

A Process of Change

For a species to adapt, a change in the environment must occur. You might think that means a change in the land or the weather, and often it does, but the environment involves much more. A mutation or gene change may occur that improves an individual's ability to find shelter. This new ability changes the environment for every other individual in the population as they all compete to find shelter. A change has taken place, natural selection proceeds, and adaptation occurs.

THE FLOW OF GENES

A mutation in one species can create a change that drives the adaptation of another species. Imagine a random mutation that might give a predator an advantage in catching prey, such as a new gene that creates a chemical to make muscles react more quickly. The quicker predator now creates a change in the environment of its prey. The prey must now adapt or it could become extinct. Any genes in the prey that would allow it to escape the faster predator would become more common in the prey's population.

Any prey without the adaptive gene would be caught, and their genes would disappear.

When genes change in a population, it is called **gene flow**. A gene may become either more or less common, depending on genetic drift or natural selection. Think of gene flow as a big lagoon, part freshwater and part saltwater. As rivers flow into the lagoon, the amount of freshwater increases. As the ocean tide rises, the amount of saltwater increases. The proportion of freshwater to saltwater changes with the seasons and the tides.

Genes do the same thing. When two populations of closely related organisms (or even closely related species) come into contact, genes might flow from one population into the other. Individuals will become more alike as genes flow in, decreasing diversity or genetic variability. In other words, as individuals from the two sides reproduce with each other, the differences between the populations would decrease. New adaptations might occur, and two separate populations (or two separate species) might eventually become one species.

For example, wolves and coyotes are classified as separate species. Normally, they would not hybridize (mate with one another) because of gene or behavioral differences. However, scientists have discovered some

DOWNLOAD

- Speciation is the process of forming new species through changes in genes.
- An isolating mechanism is any barrier that keeps two groups of organisms from exchanging genes.
- When adaptation occurs quickly after a long period of little change, it is called punctuated equilibrium.
- When adaptation occurs slowly over a long period of time, it is called gradualism.

Hundreds of millions of years ago, all of Earth's land masses were joined together in one supercontinent called Pangaea. Organisms could migrate from one area to another. As Pangaea gradually split up over time to form separate continents, animal and plant species became isolated. This isolation led to changes in genes, so new species were formed. But not all of the genes changed. Some are the same as their common ancestor's, so scientists can connect them with organisms on distant continents. For example, the North American opossum is a marsupial that carries its young in a pouch, which is evidence of its relationship to the many marsupials found in Australia.

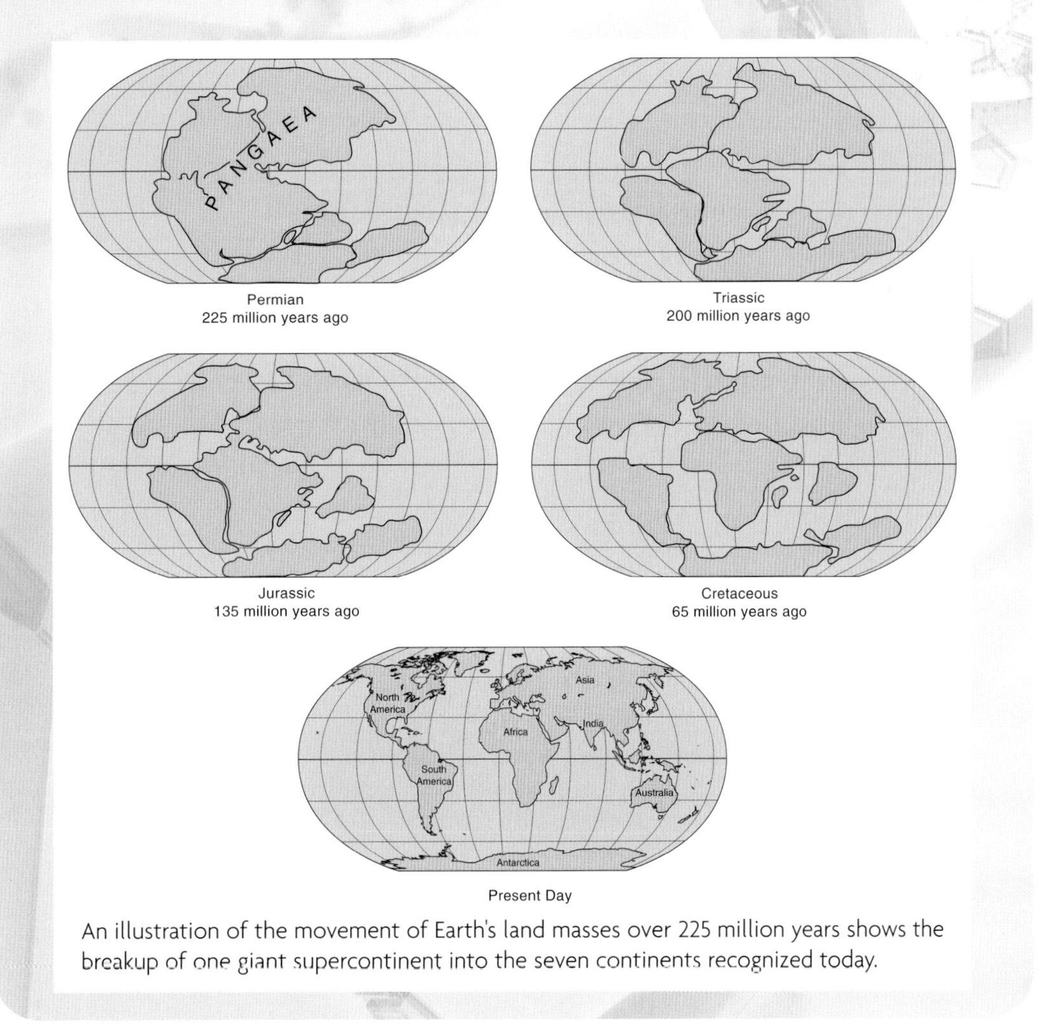

An illustration of the movement of Earth's land masses over 225 million years shows the breakup of one giant supercontinent into the seven continents recognized today.

cases of reproduction between the two. If this became common, genes would flow between the two species, and they could eventually be reclassified as one species.

On the other hand, a single population might divide to become two different species. When individuals leave a population or when genes change to make some individuals different, their family line separates or branches off from the original population. When that happens, two separate species may be formed.

The formation of species is called **speciation** *(spee-shee-AY-shun)*, which creates two groups of organisms from one. Over time, the genes of the two groups become so different that individuals from one group can no longer mate with individuals of the other group (or if they do, they cannot produce young that grow up to successfully reproduce).

DOWNLOAD

- A species is a group of organisms that can mate with one another and produce fertile offspring.
- Speciation is the separation of one population of plants or animals into two populations over hundreds, thousands, or millions of years, until they become so different that they can no longer interbreed.
- Populations can become separated or isolated by the formation of islands, new mountain ranges, canyons, or bodies of water.
- Charles Darwin observed speciation in finches on the Galapagos Islands off the west coast of South America, where the birds on separate islands differed in beak size and shape depending on the food they ate.

Anything that prevents the individuals of one population from mating with the individuals of another population is called an **isolating mechanism**. Isolating mechanisms can be physical barriers that separate populations, such as rivers, oceans, or mountains, but they can also be changes in behavior or physiology. Gene flow causes changes in the isolated populations and, once again, new species may be formed.

Isolating mechanisms can also occur within populations, creating two species that live in the same area. Even though they live together, the two groups do not mate. A change might occur in the genes of some individuals that makes their gametes incompatible with other

A mule is the offspring of a male donkey and a female horse.

individuals in the population, so that even if they did mate, fertilization would not take place. Or perhaps mating and fertilization could take place, but the developing embryo would die. This could happen because of a single difference in a gene that controls a protein needed for healthy development. The inability to create offspring could lead the two populations to become more and more different, resulting in two separate species.

Reproductive isolating mechanisms are important for many species that live together. Think of all the different kinds of trees in an area. In order to reproduce, they all release pollen, which carries the sperm of one tree to the eggs of another. The pollen from one tree may reach the eggs of a different species, but fertilization will not take place because of isolating

The courtship displays of male birds, such as the peacock, emphasize their unique colors and patterns in order to attract females of their own species.

The vocal sac of a male frog amplifies, or makes louder, its mating or advertisement call. Even when more than one species of frog is calling from the same area, a frog can recognize the unique voice of its own species.

mechanisms. Only if the right pollen reaches the right eggs will fertilization take place, creating a new tree.

Many sea creatures, such as anemones and sea urchins, release their gametes into the water. If an anemone sperm contacts a sea urchin egg, isolating mechanisms ensure nothing will happen. Only the fusing of the right sperm with the right egg will create new young. Both species can live in the same area and remain pure. Without these mechanisms, the trees would interbreed, gene flow would occur in both directions, and eventually the two would become a single species.

Other isolating mechanisms include flowers that have structures that allow only one kind of insect to enter and pollinate them. Any insects that carry pollen from different species are kept out. In animals, species may be kept separate by **hybrid infertility.** In other words, the parents from two different species may mate, but the young they produce are sterile. The young cannot mate when they are mature. Horses and donkeys, for example, produce a sterile hybrid called a mule. (Although extremely rare, a few mules have been known to reproduce.) Because their offspring are usually sterile, horses and donkeys are considered separate species.

One of the strongest isolating mechanisms between species is seen in courtship behavior. Courtship is the way a male convinces a female that he would be the best mate. A male peacock fans and vibrates its blue-green tail to attract a peahen. If another bird species with long tail feathers lives in the same area, the peahen would be able to identify the proper mate from the peacock's dance. The wrong bird just would not interest her.

Male frogs have special mating calls, depending on the species. One may sound like a foghorn, the other like a buzzer. Two or more different frog species could live in the same pond, and the females would find the right mates simply by listening to their voices.

FAST OR SLOW, CHANGE IS CHANGE

Natural selection can take millions of years to change a species. But the change can also occur very quickly. Microscopic organisms, such as bacteria, and insects reproduce rapidly, producing thousands of offspring within relatively short periods of time. Their rapid life cycle allows scientists to observe natural selection from start to finish.

When adaptations occur quickly, it is called **punctuated equilibrium**. A case in point would be bacterial resistance to antibiotics, or insect resistance to poisons. Many people are prescribed antibiotics to fight bacterial diseases. *Anti* means against, and *bio* means life, so antibiotics are capable of killing living things, specifically cells. Doctors prescribe antibiotics to kill whatever bacteria are making the patient sick. The dosage is strong enough to kill the bacterial cells, but not strong enough to kill the patient's own cells!

Doctors recommend that patients finish the prescribed amount of antibiotic, rather than taking the medicine only until they feel better. This is due to natural selection. The medicine kills many of the bacteria right away, those that lack the genes to help them resist it. But if a patient stops the medication too soon, some of the stronger bacteria, which have genes

Peppered moths rest on trees during the day where they are preyed upon by birds. Their survival depends on their genes, which determine how closely they match their background.

Staphylococcus aureus *(staff-ill-oh-KOK-us ore-EE-lus)*, commonly known as staph, is a bacterium that lives normally on the skin and in the nose of most people. It is the most common cause of skin infections and pneumonia in the United States. Normally, it can be treated with antibiotics.

However, one type of staph called MRSA (methicillin-resistant staphylococcus aureus) has become resistant to a large group of antibiotics through the process of natural selection. Dubbed the "superbug," MRSA occurs most often in people in hospitals and nursing homes who have weakened immune systems.

Bacteria can become resistant when antibiotics are not used properly. For example, some people stop taking antibiotics when they start to feel better. But at that point only the weakest bacteria have been killed; the stronger bacteria still survive. The surviving bacteria then reproduce and pass their antibiotic-resistant genes on to their offspring.

Over time, the replacement population of bacteria is made up mostly of resistant individuals. When the same antibiotic is used again on the bacteria, only a few are killed because the overall population has become resistant. That is why it is so important to take every dose of an antibiotic when it is prescribed.

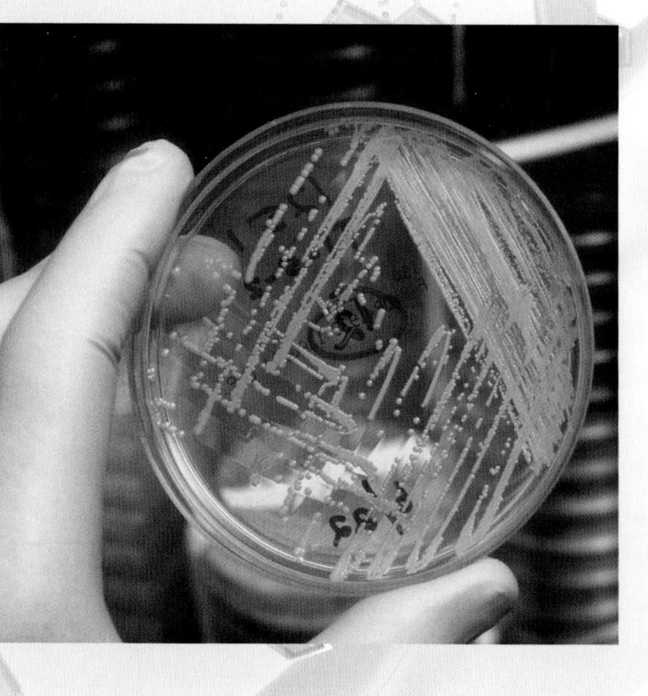

A petri dish holds antibiotic resistant Staphylococcus bacteria.

CHAPTER 3

Defenses

Every species needs to protect itself from something. Even if you live in a town or city where there are no wild **predators** to worry about, you still have adaptations that can be used to defend yourself. You have nails to scratch with, teeth to bite with, and hands that fold into fists to hit with. If a threat is too great to face, your leg muscles are well adapted for running away. These ways of defending ourselves from danger are known as fight-or-flight responses.

Natural weapons, such as claws and fangs, are common among living things, but they are not the only defensive adaptations. Defensive adaptations can be behavioral, chemical, or mechanical. A behavioral adaptation has to do with the way an animal acts. It might hiss, growl, or act scarier than it really is. It might go underground or climb a tree. Chemical adaptations include tasting bad, squirting something nasty, or producing a toxin or venom.

Mechanical adaptations have to do with body characteristics or structures on the body that help a species protect itself. Keep in mind that an adaptation is a characteristic that is controlled by a gene or genes. The genes make proteins that are used by cells to create characteristics or

A porcupine's quills are a mechanical adaptation that protects it from attack. Some predators, however, such as the lion, have learned to flip the porcupine over to get at its soft underside *(upper left)*. Growling—such as the wolf is doing—hissing, and spitting are defensive adaptations used by a variety of animals to warn others to stay clear *(above)*. A toad's poison glands (large bumps behind the eyes) are a chemical adaptation. The poison it excretes causes burning and irritation in the mouth of any attacker that grabs it *(lower left)*.

adaptations. **Mimicry** is the way in which living things imitate other natural objects, either living or nonliving. In other words, they look like something that they are not.

Mimicry is a mechanical adaptation because it involves structures of the body, such as skin. Camouflage is a special kind of mimicry in which an organism blends into its natural environment. It might look like a rock, dirt, or a part of a plant, such as a leaf or a twig. A speckled snake that looks like gravel is an example of camouflage. A fly that imitates a bee is not an example of camouflage because the fly would stand out against its background.

MASTERS OF DISGUISE

Prey animals use camouflage to hide from or confuse predators. They can then hide in plain sight. Hiding also helps an animal to conserve energy that would otherwise be spent fighting for its life or running away. Prey animals with the best camouflage genes are more likely to survive predation and pass those genes on to the next generation.

Camouflage also helps predators hide in ambush. Predators save energy by not having to search for their food, because the food comes to them. Predators with the best camouflage genes catch more prey and are more likely to survive to pass those genes on to the next generation. This battle between the best genes in predator and prey is played out every single day in the struggle for survival in the natural world.

Camouflage works by changing or disrupting the normal form, or **gestalt** (*geh-STAHLT*), of an animal. Gestalt is the total sum of an individual's characteristics that are recognized together as a whole. Change one part of the whole and the individual becomes harder to recognize.

Sharks have a type of camouflage called countershading. In countershading, the top of the body is dark and the bottom of the body is light. It is common in many ocean fishes, but also in aquatic animals like whales, penguins, and frogs. When seen from the top, the dark back blends into the darkness of deeper water. When seen from below, the light belly blends into the sunlit surface of the water. Countershading conceals predator from prey, and hides prey from predator.

Countershading in sharks, as well as in frogs, whales, and other fishes, camouflages them from both above and below.

Cryptic coloration and shape in a praying mantis make it hard to tell the animal from the plant it rests on.

A man who wears a beard for years and then shaves it off or a woman with blonde hair who dyes it black would be difficult to recognize at first. They have changed the gestalt pattern that the observer's brain would normally recognize.

Predators have two types of ability that allow them to recognize the gestalt of their prey: innate or inborn, and learned. The predator recognizes the shape, body characteristics, and movements of its prey. Prey animals recognize predators the same way. If individuals inherit genes that change the normal gestalt of the species, they are more likely to avoid predation. If predators inherit **sensory** genes, which intensify their senses, they are more likely to find camouflaged prey.

Animals may be camouflaged from head to toe, or only a part of their body may be affected. They look like things that would occur in their natural environment. They might resemble living leaves if they live in trees, dead leaves if they live on the forest floor, mud if they live near water, or sand if they live in a desert. This type of camouflage is also called **cryptic coloration**. Any inherited genes or mutations that create characteristics that do *not* match objects in the background would make the animal stand out and have the opposite effect. Prey would be picked off more easily, and predators would not be as successful at sneaking up on their prey. Through natural selection, these unhelpful genes would die off with the unfortunate animals.

- Mimicry is the way in which living things resemble objects in nature, which gives them a survival advantage.
- Camouflage or cryptic coloration is a type of mimicry that involves blending into the environment.
- Warning coloration signals that a plant or animal has a bad taste or is poisonous.
- The model is the object that is mimicked; the mimic is the individual that looks like the object.

HIDING AMONG THE LEAVES

Plant mimics are common in nature and occur around the world. Phylliid leaf insects (*phyllas* is Greek for leaf) of Asia and Australia are green and flat. Not only are their bodies leaf-shaped, but flattened green flaps extend out from the sides of their legs so that each leg looks like a leaf. When these insects cling to trees, they look like another bunch of leaves growing out from a stem.

The Madagascan katydid mimics not only the green color of a leaf, but also its speckles and veins. The outside edge of the katydid's body is even ringed with tiny thorns that look like the serrated edges of some leaves.

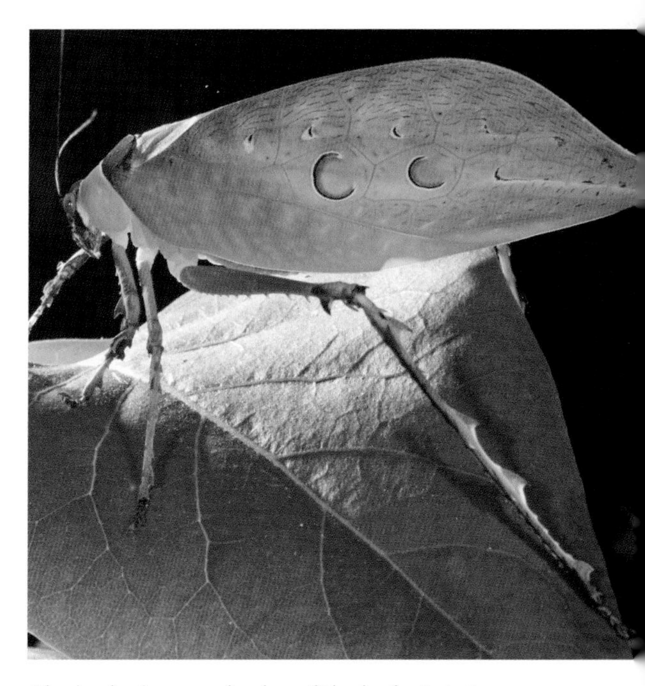

The body shape and color of the leaf mimic is a near perfect imitation of a real leaf.

Many frogs are also green and spotted like leaves. When perched on a leaf, they tuck in their legs and disappear against the leaf's surface.

Amphibians, fish, reptiles, and birds have an extra clear eyelid, called the third eyelid or nictitating *(NIK-teh-tay-ting)* membrane, which slides sideways across the eye to keep it clean, moist, and protected. However, the third eyelid of the Central American red-eyed tree frog is also streaked with golden lines that look like the veins of a leaf. When the green frog tucks itself in against a leaf, it can close the third eyelids while keeping an eye out for danger through the clear spaces between the lines. (Humans also have a third eyelid in the inner corner of the eye, but it is vestigial and is so small that it can barely be seen.)

Many animals, such as the praying mantis, mimic dead leaves rather than live green ones. The Indian leaf mimic butterfly is glittery blue with yellow stripes when its wings are spread for flight, but when it folds them together up over its back, it looks like a dead brown leaf. The Australian spiny leaf insect, besides mimicking the golden brown of drying leaves, has spiny projections on its body. If its camouflage fails it, the spines may prick a predator's tongue, causing it to spit the insect out. Many plants use spines and thorns as a means of defense. These pointy projections keep animals from eating their leaves.

Many ground-dwelling animals, such as frogs and snakes, are blotched, speckled, or mottled like the fallen leaf litter that covers the forest floor. Their bodies may also be leaf-shaped. The Asiatic horned frog is flattened, with a pointed nose like the tip of a leaf, projections over its eyes that give it a jagged look, and folds down its back that look like leaf veins.

Some birds mimic tall reeds and grasses, and some fish mimic seaweed. A bird called the bittern lives among the reeds of North American marshes. It has long legs, neck, and beak, and feathers that are streaked with black to blend in with the shadows among the reeds. When the bittern senses danger it freezes, stiffens its long neck, and points its beak to the sky. It even sways in the wind with the reeds. The pipefish looks exactly like a long, thin piece of seaweed. Often, a group of pipefish huddle together, swaying with the currents like the undulating underwater plants.

STICKS AND STONES

There are animals from all over the world that look like twigs or branches, including stick bugs, walking sticks, vipers, boas, and vine snakes. These animals may be gray, brown, striped, or speckled to blend in with the bushes and trees. The eggs of many insects, such as the katydid, look like the scales found on branches or twigs.

Many lizards, moths, and other insects look like tree bark. Their bodies are often ragged around the edges with flaps of skin or spines that further blur their outline and make them harder to see. The Australian leaf-tailed gecko looks so much like the roughened bark of trees that it is hard to see even when you know where it is. Some moths hide against bark, but if detected or disturbed, they do not fly away. Instead they fall straight to the ground like a dead leaf.

Giant walking stick insects have specialized legs and coloration that mimic the structure of the tall grass on which they are perched.

The slender vine snake has adapted to mimic the tree branches and vines of its natural habitat.

Even more bizarre are treehoppers and thorn bugs. Both look exactly like the thorns that are found on many bushes and trees, except these thorns have legs and can get up and walk away! Many spiders are well camouflaged for tree life, but the yellow crab spider takes the cake. It blends perfectly into brightly colored yellow flowers so that it is protected from predators. The yellow crab spider is also a predator itself. It waits for an unsuspecting bee to be attracted to the flower, and when a bee arrives seeking pollen, the spider captures and eats the bee.

Animals of all kinds mimic the earth, from dirt, sand, and rocks to slimy muck. Spotted fawns, left behind when their mothers move off to feed, are hidden against the ground. Squat North American horned lizards and scaly sidewinders blend into pebbly sand. The rockfish looks exactly like a jagged underwater rock. The bumpy scales on an alligator's snout and back give the illusion of a floating log as it sneaks up on its prey. Alligators, turtles, and crabs are all able to match their aquatic environments, and many have additional camouflage that is not determined by their genes: algae grow on their backs or shells to complete their disguises.

An alligator blends into its surroundings while lurking in pond water. The alligator uses this ability to surprise and attack its unsuspecting prey.

The North American hognose snake relies on its dull coloration to hide it against the ground, but it also has a backup plan. If detected and threatened, it rolls over on its back with its tongue lolling out of its mouth and imitates a dead snake. Many predators are only attracted to moving prey, and when their target does not move, they lose interest. When the hognose feels safe again, it rolls back over and slithers away.

The yellow and black stripes on a Bengal tiger allow it to blend into the light and shadows of tall grasses and brush.

Many predators are stimulated to attack by movement in their prey. When a hognose snake freezes and plays dead, it may avoid being eaten.

Many animals do not mimic any particular object, but their bodies are spotted or striped to blend in with the shadows that surround them. Leopards, tigers, and zebras are good examples. During the day, zebras may seem conspicuous, but they are more likely to be hunted as the sun goes down. In the dusk, their stripes blend into the shadows of tall grasses and bushes, which makes it more difficult for predators to see them.

The zebra's camouflage is called **disruptive coloration** because the stripes disrupt or interrupt a predator's image of what prey is supposed to look like. Eye stripes, which are seen in everything from fish, lizards, and snakes to antelope and wildcats, are another type of disruptive coloration. Predators have trouble telling where the head end of their prey is, and prey cannot see the eyes of the hunter sneaking up on them.

Some animals even have the genetic ability to change their color to match the background. Tree frogs transform slowly, changing from the bright green color of new leaves or mosses to the dark browns of mud or

See if you can find the animal camouflaged in each picture.

A lizard camouflaged on a tree branch *(left)*, a seahorse camouflaged among polyps *(center)*, and a gecko camouflaged on a tree trunk *(right)*.

bark. Cuttlefish and chameleons have the ability to change almost instantaneously as they move from one environment to another. Color changes reflect the mood of these animals. When frightened, they turn dark and mottled in an attempt to disappear. When in the mood for love, the cuttlefish blushes red-orange to advertise its beauty and attract a mate. Depending on the species, the chameleon flushes bright green with blue, yellow, white, or red streaks and spots.

Other animals change with the seasons. The snowshoe hare changes from dirt brown in the summer to snow white in the winter. All of the color changes in animals are manipulated by chemicals controlled by the animal's genes.

While animals are often adapted for camouflage, plants rarely are. One exception is the stone plant, which lives in the African desert. The juicy leaves of the stone plant hold water to help it survive in its hot, dry environment, but they also make these plants an ideal food and water source for desert animals. Certain genes were favored over time resulting in the stone plant's resembling a cluster of stones, effectively blending the plant into the desert background.

Stone plants merge with their desert background.

WATCH OUT!

Instead of hiding through camouflage, some animals do just the opposite. They show off or advertise their bright colors. The red, orange, yellow, and blue found on various types of South American poison arrow frogs warn predators that they are poisonous. If an inexperienced predator eats one, glands in the frog's skin exude a bitter-tasting toxin that will make the predator spit the frog out. Later, the predator associates the bad taste with the frog's bright color and looks for a meal elsewhere.

Warning coloration is found in many species of frogs, poisonous Gila monsters, venomous coral snakes, stinging bees and wasps, and in many beetles, butterflies, and ants. The poisons produced by these animals are strong enough to harm or injure a predator, but normally will not kill it. A dead predator cannot remember the experience and avoid the prey in the future, so the prey would not receive a survival benefit. Instead of being

Watch out for me, or you will regret it! Many poisonous animals advertise their bad taste or venom. It only takes one mistake for a predator to remember which animals to avoid in the future.

Many venomous animals—such as the Gila monster *(above)*, coral snake *(upper right)*, and wasp *(lower right)*—are marked with bright warning colors.

avoided, they would continue to be preyed upon, and the gene for the stronger poison would die out. On the other hand, the gene that controls the poison would have to be strong enough to make a lasting impression, or it, too, would die out.

Some animals rely on camouflage first, but if that fails, they have a backup plan. Fire-bellied toads, found in Europe and Asia, are camouflaged on top to look like green algae and are bright red or orange underneath. If their camouflage fails them and they are threatened, they arch their heads and flash their colored bellies to remind predators that they taste really bad.

A poison is any chemical substance that is capable of injuring or killing an organism. It may be natural or manmade. Toxins are poisons that are produced naturally by living organisms. For example, toadstools produce toxins that are poisonous to eat. Toads excrete foul-tasting toxins from special glands behind their eyes so predators will spit them out.

Animal toxins that are delivered by a sting or a bite are called venoms. Venomous animals produce their venom in order to feed or to defend themselves. Vipers, cobras, scorpions, ants, and bees are venomous. Bees sting to defend their hives. Venomous snakes inject their venom to immobilize their prey, which comes in handy for an animal without hands.

Venom varies in the way it affects body tissues. Some venoms attack the nervous system, causing respiratory or heart failure. Others cause internal bleeding, keeping blood from clotting normally. Still others break down and destroy muscle tissues, much like a stomach digests food, so the victim cannot move or get away. In fact, the glands that make a snake's venom are modified salivary glands and venom is modified saliva. Like venom, the saliva in your mouth begins the digestive process even before you swallow.

Venom is delivered through hypodermic needle–like fangs when a rattlesnake bites.

The green color of the fire-bellied toad camouflages it when floating in a mossy pond *(left)*. If the camouflage fails it, the toad flashes its red-orange belly to warn of its bad taste *(right)*.

The Australian shingleback—or two-headed skink—has a tail that is camouflaged to look like its head. If a predator takes a chunk out of the skink's tail, the skink has a chance to get away and survive. If the predator goes for the skink's head, it is another story. If the skinks' trickery fails them, they open their mouths and stick out a bluish black tongue. The sudden gaping of a large mouth and the flashing of the tongue is often enough to scare an enemy away. An adaptation that is hidden most of the time and flashed only under dire circumstances is called **flash coloration**. Flash coloration, along with gaping, puffing up, hissing, and screaming, are tactics that make an individual look bigger or seem more frightening than it really is. Basically, it is a fake-out.

Many moths and butterflies have eye-like spots on their lower wings that are hidden when the upper wings cover them in camouflage. When needed, they can flash the fake eyes at birds, their main predators. Birds fear the eyes of their predators, so the sudden appearance of large colorful eyes can be enough to startle them away. Pygmy owls have false eyes on the backs of their heads to discourage predators from sneaking up from behind.

A shingleback lizard gapes and sticks out its tongue in order to look fiercer than it really is.

When threatened, hawk moth caterpillars swell up their behinds to reveal startling blue eyes, so that they resemble a big snake. The Chilean four-eyed frog has its normal eyes in front, but also a bright pair of fake eyes on its hind end. When the frog is sitting on the ground, the extra "eyes" may be partly or totally covered by the hind legs. But if threatened, the frog tucks down its head and raises its rear end to flash the scary eyes.

Then there are the fakers that imitate other dangerous animals. Because predators learn to avoid brightly colored animals that may be poisonous, they also avoid any brightly colored animals that are not poisonous. A number of butterflies mimic others that taste bad. The edible Indian fritillary butterfly is such a close mimic of the unpalatable, or bad

The viceroy butterfly was once thought to be a mimic of the foul-tasting monarch butterfly. But recent studies have shown that the viceroy is just as poisonous as the monarch. Scientists now consider them to be examples of a totally different kind of mimicry, where two toxic species mimic each other. Both butterflies benefit because any predator that encounters one of them will also avoid the other.

Predators that try to eat the foul-tasting viceroy butterfly learn to avoid them by associating their bad taste with their color.

tasting, plain tiger butterfly that an expert is needed to tell them apart. Interestingly, the bad taste of toxic butterflies is not controlled by genes. Instead, it comes from chemicals in the leaves of plants that they eat when they are caterpillars.

The leaves and stems of poison oak produce a toxic oil, which can cause a painful skin rash in people who come into contact with it.

The dangerous animal that is being mimicked is called the model. The harmless faker is called the mimic. Models may have one or many mimics. Stinging bees and wasps are models for a number of mimics, including flies and moths. For example, the drone fly mimics the honeybee, and the ash borer moth mimics a common wasp.

Many harmless snakes mimic venomous snakes. Harmless milk snakes mimic the bright yellow and red bands of the venomous coral snake. A number of harmless ground snakes rattle their tails like rattlesnakes. The tail rustles dried leaves on the ground, creating the rattlesnake sound. The rattle of the rattlesnake itself is in fact an adaptation for defense. The sound is enough to warn away large hoofed animals, which keeps the snake from getting stepped on. The rattlesnake's venom is not a defensive adaptation, but one that it uses to subdue prey. When they bite in defense, they often do not even inject their venom.

Poisons are a form of chemical defense and are common in both the plant and animal world. If the leaves of a plant taste bad, then they will not be eaten or trampled on by browsers. The leaves of the pencil tree and many other plants exude bitter white goo when eaten. Toxic plants like poison oak and poison ivy are colored red in the autumn, a form of advertising color that protects the plant during a time when it is not growing.

Some plants even give off a chemical odor when insects chew on their leaves. The chemical attracts predatory wasps, which fly to the plant and eat the harmful insects. The plant lives. And do not forget the skunk. Specially adapted glands under its tail spray a horrific burning liquid at whatever threatens it.

When it comes to defense, whatever works is the name of the game.

Brown bears learn to wait
at the tops of waterfalls
for salmon migrating
upstream to lay their eggs.

CHAPTER 4

Feeding

Living things are defined by the common activities that they all carry out. They must take in nutrition and **respire** to get the energy out of their food. They must excrete wastes, respond to changes in the environment, grow and develop, and reproduce. All of these activities are controlled by genes. A disruption in any one of them can lead to the death of an individual or the extinction of a species.

In order to carry out the activities of life, living things must have access to certain necessities, such as oxygen, light, water, and food. Because food is the most important source of energy for living things, each species has many different adaptations for obtaining it. Whether or not an organism has a mouth, it has to have some way of getting nutrition into its cells and then making use of it.

Single-celled bacteria have genes that excrete chemicals to break down the plant or animal cells upon which they feed. They must also have adaptations for locating their food, processing it, and getting rid of the wastes that are left over. People have hands to grab food, lips to move the food into their mouths, tongues to move the food around in the mouth, teeth to chew the food and break it down into smaller pieces, salivary glands to moisten

Baboons and other primates, including humans, have hands that are adapted for finding, picking at, grabbing, and holding their food *(above)*. Without hands, snakes must use their bodies to grab their food *(left)*.

the food, throat muscles to squeeze the food down toward the stomach, and so on. An animal that does not have hands will have other adaptations to take their place. A snake, for example, grabs food with its sharp teeth, or it might use its entire muscular body as one giant hand.

MANY WAYS TO EAT

Most animals have to go out and find their food, but a plant's food has to come to the plant. Through a process called **photosynthesis**, plants make their own food in the form of sugar using the light from the sun. There are exceptions, of course. **Carnivorous** plants, such as the Venus flytrap and the pitcher plant, lure insects into their specially adapted leaves, trap them, and then digest them. These plants still use photosynthesis to harness the energy of the sun, but the insects provide them with additional nutrients. A number of gene adaptations control the growth of leaf forms that trap insects and the production of the proper enzymes for digesting the insects they catch.

Animals might hunt, stalk, ambush, or even lure their prey. The adaptations that help them are related to the type of food they eat. Animals that

More than 600 species of plants have adapted over time through natural selection to become carnivorous. The specialized leaves of the Venus flytrap look like a set of miniature jaws with long jagged teeth that snap closed when touched. Any insect that steps on the jaws is trapped inside.

Pitcher plants and lobster pot traps have vase-shaped leaves that form slippery chambers that are easy to enter but difficult to leave. Some have spiky hairs inside that point downward, preventing any insects from climbing out once they have been trapped inside the cup. The bottom of the cup is filled with digestive enzymes secreted by the plant, similar to the ones people secrete into their stomach and intestines. Sundew traps have long, tentacle-like leaves that secrete a sticky liquid. Any insects that land on them are glued in place. The leaf then curls over the insect and digests it.

Carnivorous plants catch and devour prey. A wasp is caught by a Venus flytrap *(left)*. A pitcher plant catches and digests insects in a toilet bowl–like cup *(right)*.

burrow for earthworms might have short necks to push through dirt or long beaks for poking into the earth. Animals that browse in trees might have long necks for reaching low branches or gripping feet for climbing up into the branches.

An animal's adaptations allow it to find food and to deal with it once it is found. Animals that chew leaves have flat teeth, while those that eat meat have cutting teeth. Animals that eat slippery fish have curved, puncturing teeth so that fish cannot slip away. Birds do not have teeth. Instead, they swallow their food whole or use their beaks to break it apart.

Chameleons have sticky, spring-loaded tongues that shoot out and capture insects from a distance. Giraffes have long, flexible tongues for ripping the leaves off branches. African clawed frogs and many aquatic salamanders, such as the Mexican axolotl (*AK-suh-laht-uhl*), have an immovable tongue that they do not use at all. They simply suck in their food like a vacuum

To catch insects, a chameleon can shoot out its sticky, spring-like tongue one and a half times the length of its body.

The flexible jaws of the egg-eating snake allow it to engulf eggs much larger than its own body. Special bones in its neck then break the shell so the snake can swallow the egg's contents.

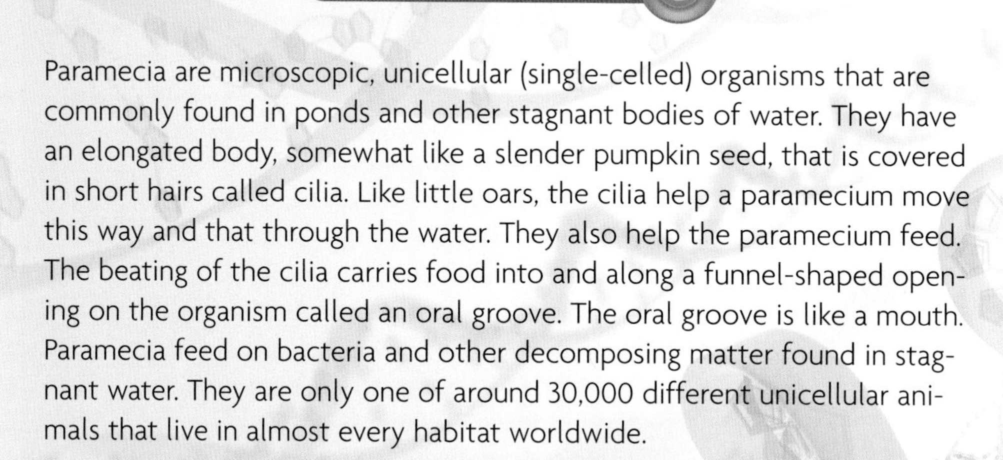

POP-UP

Paramecia are microscopic, unicellular (single-celled) organisms that are commonly found in ponds and other stagnant bodies of water. They have an elongated body, somewhat like a slender pumpkin seed, that is covered in short hairs called cilia. Like little oars, the cilia help a paramecium move this way and that through the water. They also help the paramecium feed. The beating of the cilia carries food into and along a funnel-shaped opening on the organism called an oral groove. The oral groove is like a mouth. Paramecia feed on bacteria and other decomposing matter found in stagnant water. They are only one of around 30,000 different unicellular animals that live in almost every habitat worldwide.

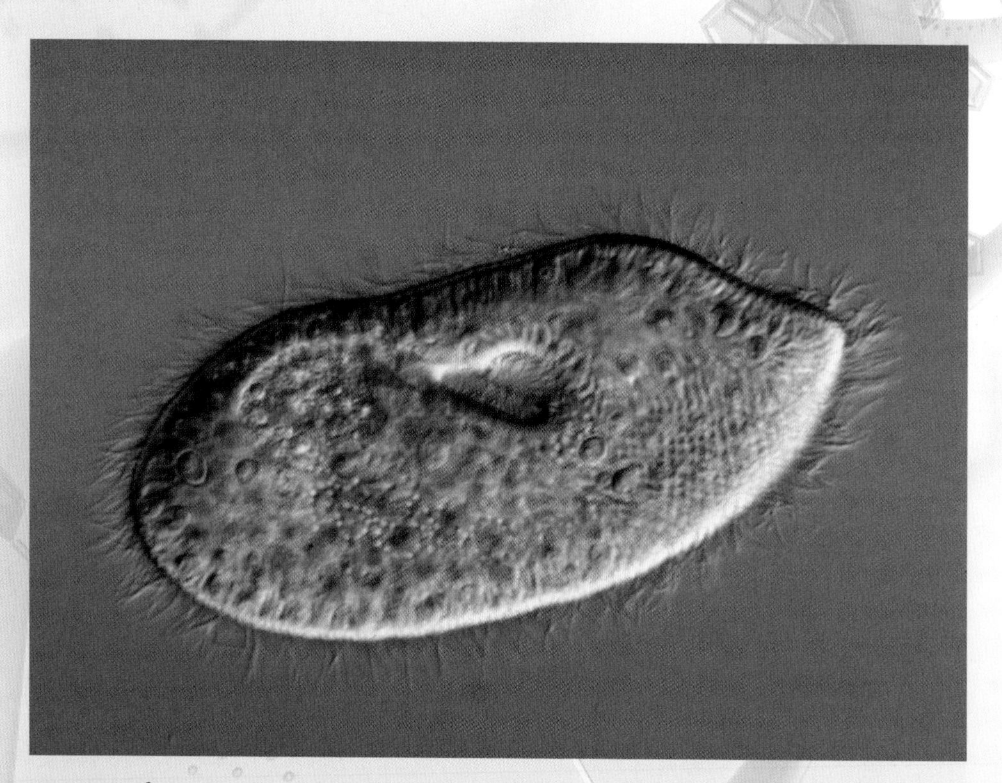

An image of a *paramecium* taken with an electron microscope clearly shows the hair-like cilia that surround the organism.

The forked tongue of snakes is actually an organ of smell. Odors are transferred from the tips of the tongue to the nose through special organs in the roof of the snake's mouth.

cleaner. Egg-eating snakes have a special bone that protrudes into the throat and breaks the egg as it goes down.

Rattlesnakes and other pit vipers have special adaptations that help them find their prey, even in total darkness. A pit is located on each side of the head, between the nostril and the eye. The pits contain specialized nerves that act as infrared heat sensors. The nerves are sensitive enough to detect temperature differences as little as 0.20° C. Messages from the nerves send a heat picture to the snake's brain, allowing it to "see" the warm body of a mouse (or anything else that gives off heat) as it moves around at night.

Once a rattlesnake, or any poisonous snake, locates its food, it uses venom to subdue it. Snake venom is saliva that has been modified by changes in the snake's genes. The two canine teeth of venomous snakes have also been modified through gene changes to form fangs. Each fang is needle-shaped and hollow. One quick bite and the venom is injected, as easily as a doctor giving a vaccination.

If the prey runs away after a bite, the snake simply follows. Thanks to a specially adapted tongue, a snake does not need to see where its meal went. A snake's forked tongue is a powerful organ of smell, which it can use to follow the scent trail left behind by even the tiniest footsteps. Once the venom takes effect, the bitten animal collapses and the snake follows its scent trail.

Boas and pythons have many infrared heat-sensing pits that run along the tops of their lips. Their pits allow them to track their warm-blooded prey, including birds in trees, at night. Boas, pythons, and vipers sense prey in three ways: heat-sensing pits, tongues that smell, and **elliptical** pupils in their eyes for better night vision.

Elliptical pupils look like narrow slits. Nocturnal animals, such as cats, alligators, frogs, and geckos, have them. The narrow shape allows more light to enter the eye at night. Round pupils, such as those in human eyes, are adaptations for seeing in daylight.

Animals with vestigial eyes that live in perpetually dark places, such as caves and the leaf litter of dark forest floors, sense changes in light intensity, but they cannot see shapes. Producing body parts that an individual does not need, such as eyes underground, would be a waste of energy. So natural selection tends to weed out the genes that produce them. Wasted energy can weaken an individual and make it less likely that it will survive to pass on those unneeded genes.

Senses such as smell become more important for finding prey in dark places. The North American garter snake hunts for its food, such as frogs and salamanders, under dark logs and decaying leaf litter around ponds. The snakes have sensitive eyes, but can find food by smell alone. Their prey, however, have adaptations to avoid consumption. The California slender salamander thwarts the garter snake by looping its tail around the snake's head and neck, making it impossible for the snake to swallow. During the struggle, the salamander secretes a sticky substance from specially adapted glands in its skin. The snake ends up with its body glued to itself and its mouth glued open. The salamander releases its grip and escapes.

DEADLY DECEPTIONS

Sharp teeth, claws, and talons are common adaptations for hunting. Less familiar are the adaptations used for luring prey. Animals that lure mimic the natural food, or even the mates, of the prey they eat. They do not search for their food; the food comes to them. Carnivorous plants also lure food, usually with sweet nectar as the bait.

Many animals are masters at luring. The Mexican cantil (a venomous pit viper), the Australian death adder, and the New Guinea tree python all

The alligator snapping turtle has a pink, wormlike extension of its tongue that it uses to lure fish. Special muscles in its neck are able to constrict suddenly, causing a sucking motion that sweeps the fish into the turtle's mouth. Although the turtle has no teeth, it has a beak and powerful jaw muscles that can crush even the largest fish.

A pink, worm-like appendage in the mouth of an alligator snapping turtle lures fish into the turtle's mouth.

use their tail to entice their prey. The bodies of the snakes are camouflaged to blend into the background so their prey cannot see them. The vipers are dark like the ground and the tree python is green like the trees. The tips of their tails, however, are shaped and sometimes colored like worms. The snake coils up with its tail close to its mouth. Then it wriggles its tail in perfect mimicry of a squirming worm. Small mammals and birds are attracted to the phony worm. Bam! The snake strikes and dinner is served.

The firefly is actually a special kind of beetle. The light, produced by a chemical reaction in its lower abdomen, is used to attract mates or prey. Depending on the species, the color of the light may be green, red, or yellow, as in the species shown at left.

Alligator snapping turtles use the same technique, except their lure is formed by a special worm-like extension of their tongue. The camouflaged, mud-colored turtle rests on the bottom of a pond with its mouth open. Wriggling the mock worm, it simply waits until a fish is attracted to the lure. When the fish gets close enough, the turtle simply swallows the fish down with a sudden sucking motion of its throat.

The South American horned frog uses its toes to lure its prey. This **terrestrial** frog burrows down into the soil and leaf litter, raising a rear toe up over its back. Then it wiggles its toe like a worm. It has a huge mouth and will eat anything that approaches its lure, even another horned frog!

In the eastern United States, one species of firefly mimics the flashing lights of the female of another type of firefly. When a male from the other species is attracted to the female mimic, the imposter captures and eats the male. Nocturnal fireflies have organs in their abdomens that give off the glowing light, thus using **bioluminescence** *(by-oh-loo-muh-NESS-ehnts)*. Special genes cause a chemical reaction inside the organs that creates the glow. Fireflies that are active during the day do not have the same adaptation.

YOU SCRATCH MY BACK, I'LL SCRATCH YOURS

Symbiosis *(SIM-bee-OH-sihs)* is a relationship between two different kinds of organisms in which one or both species profit in some way. There are three forms of symbiosis: mutualism, parasitism, and commensalism.

In **mutualism** *(MEWCH-oo-uh-liz-uhm)* both species benefit from the relationship. Mutualism exists between rhinoceroses and the birds that clean them, and between ants and aphids. The birds, called oxpeckers (they also peck on cattle), feed on ticks and fly larvae, receiving a meal for themselves while removing the harmful parasites from the rhinos. Ants feed on the sugary solution excreted from plant-sucking aphids, and the aphids are protected by the stinging ants.

The lichens that grow on trees and rocks are formed by fungi and algae that live close to each other. The fungi absorb minerals from the rocks and bark that the algae can then use to carry out photosynthesis. In return, the algae provide the fungi, which cannot photosynthesize, with food.

The feeding association between a small ocean fish called the cleaner wrasse and its host fish is also type of mutualism. Larger fish recognize the wrasse by its coloration and its dance. The dance convinces the larger fish to open its mouth, but not to eat the wrasse. The wrasse then cleans up any parasites around the larger fish's mouth, body, and fins. The wrasse gets a meal, and the host is cleaned of its parasites.

Another small fish, called the blenny, mimics the wrasse. When a larger fish allows the imposter to approach, the blenny takes a bite out of it and streaks away. In that case, the blenny benefits, but the host fish is harmed. This form of symbiosis is called **parasitism** *(PA-ruh-sih-ti-zuhm)*. Parasites cannot survive on their own without their host and have special body adaptations that help them maintain their relationship. Normally, the host is sickened or weakened in some way, but not killed. Natural selection has favored this, because if the parasite kills the host, it kills itself and its genes are lost.

Mosquitoes, ticks, and fleas are well-known parasites. The mosquito has special organs that sense the carbon dioxide that is exhaled by its host. Its syringe-like mouthparts are adapted for sucking blood. Ticks and fleas sense body heat and have bristly legs that help them grip the skin of their moving

An oxpecker feeds on parasites on the skin of an African Cape buffalo, a wild member of the bovid (cow) family.

hosts. All of these animals are called **ectoparasites** because they live on the outside of the host's body. *Ectos* is the Greek word for out.

Endoparasites live inside a host's body. *Endon* is the Greek word for within. Sleeping sickness is caused by a single-celled parasite that enters the bloodstream through the bite of the tsetse fly. Roundworms and tapeworms are common parasites that live in the intestines of dogs and other animals, including humans. Thousands of endoparasites are known to cause disease.

Large numbers of heartworms are able to live and feed in the hearts of dogs, coyotes, wolves, and foxes, and can infect other animals, such as cats and humans. Heartworm infestation is easily treated, but if left untreated can eventually cause heart failure. Heartworms give birth to live young that circulate in the host's bloodstream. When the host is bitten by a mosquito, the heartworm larvae enter the mosquito. When the infected mosquito feeds again, the larvae are passed on to the next victim, and the cycle continues.

A number of plants are parasitic. Mistletoe, which is so often hung above doorways at Christmastime, is a parasite on trees. Its roots pierce the tree's bark and invade its branches, stealing nutrition and stunting the tree's growth. The strangler fig, found in many tropical forests, starts as a seed that is carried to the top of another tree by a bird. When the seed germinates, it sends roots through the air down toward the ground, taking nutri-

DOWNLOAD

- Symbiosis is a relationship in which two different organisms live together and one or both benefit.
- The three basic forms of symbiosis are mutualism, parasitism, and commensalism.
- Mutualism is a relationship between two different organisms in which both benefit.
- Parasitism is a relationship between two different organisms in which one benefits and the other is harmed.
- Commensalism is a relationship between two different organisms in which one benefits and the other is unaffected.

ents from the host tree. Eventually the strangler fig roots itself into the ground and no longer needs the tree, but by that time it has engulfed and killed its host, leaving a huge hollow space in its center.

Commensalism *(kuh-MEHNT-suh-liz-uhm)* is another form of symbiosis in which one organism benefits from the relationship, but the host is neither helped nor harmed. Tiny ocean shrimp often live and feed on other sea creatures, such as anemones and sea slugs. Clown fish live among the stinging tentacles of sea anemones. The clown fish, which is immune to the anemone's sting, benefits by being protected from predators.

A fish called the remora has a modified dorsal fin that acts like a suction cup to attach the fish to a shark. The shark is not harmed, but the remora receives a free ride, bits of food left over from when the shark feeds, and protection from its enemies.

Tropical American bromeliads *(broh-MEE-lee-ahdz)*, relatives of the pineapple, live on the branches of trees. They benefit from the shelter of the tree where they receive more light and air than they would on the ground, but the tree neither gains nor loses from the presence of the bromeliad.

A clown fish is unharmed by the stings of a sea anemone.

Adaptations are easily recognized when they are at the tissue or organ level: skin, feathers, scales, horns, and so on. However, all adaptations originate at the cellular level where the DNA is housed. The genes inside those cells are what signal the formation of body structures as well as the orchestration of bodily functions, such as digestion to break down food or excretion to get rid of waste. In a sense, all living things are adapted to feed at the cellular level.

Sexual reproduction involves the creation of a unique individual from two different parents.

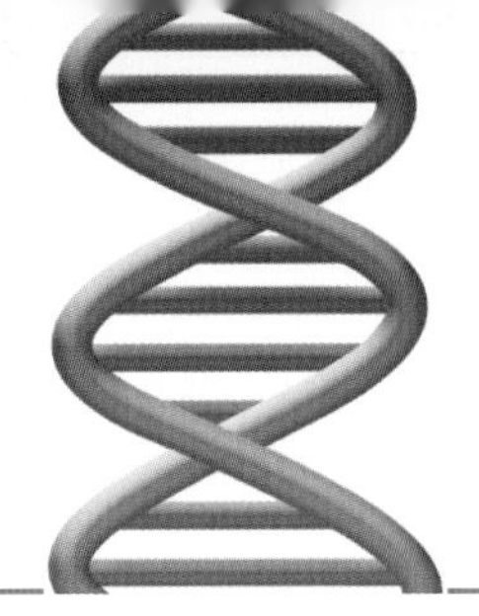

CHAPTER 5

Reproduction

Have you ever had a litter of puppies or kittens in your home? Have you ever kept chickens in order to collect their eggs? Or planted a vegetable garden by sowing the seeds? Most people are familiar with the common ways in which plants and animals reproduce. Not quite as familiar is the great variety of adaptations that different species use to ensure their genes are passed on to the next generation.

PASS IT ON

Two forms of reproduction are used by living things: asexual and sexual. **Asexual reproduction** requires only one parent. The parent simply makes a copy of itself. Because asexual reproduction does not involve **fertilization** of an egg by a sperm, it has the advantage of being simpler and faster than sexual reproduction. Most single-celled organisms, such as bacteria, as well as body cells in plants and animals, reproduce asexually.

Sexual reproduction involves two parents, a male that produces sperm and a female that produces eggs. Each sperm carries half the genes of the male. Each egg carries half the genes of the female. When fertilization takes place, a new individual is created with all of the genes it needs to survive. The offspring are not exact copies of either parent, but a combination of the characteristics of both parents. Sexual reproduction may not be as easy or fast as asexual reproduction, but it has the advantage of increasing genetic diversity. Genetic diversity increases a species' ability to adapt to changes in the environment.

Sexual reproduction is a challenging process. Plants cannot move, so they have adaptations that get the male spores or pollen to the female plant, using the wind, gravity, water, insects, and animals as their moving companies. Each method requires different adaptations. Even if a plant **self-pollinates**, the gametes, or sex cells, still have to get from one part of the plant to another. Often, plants have colorful flowers that attract insects. The insects then carry the pollen from flower to flower. There is even an orchid with flowers that mimic a female wasp. Male wasps are attracted to one orchid, pick up its pollen, and carry it to the next fake female on another orchid. By fooling the wasp, the orchid is pollinated.

Animals must find or attract a mate, and competition is often stiff. In most cases, success depends on having gene adaptations that make the suitor more attractive in some way: bigger, stronger, more

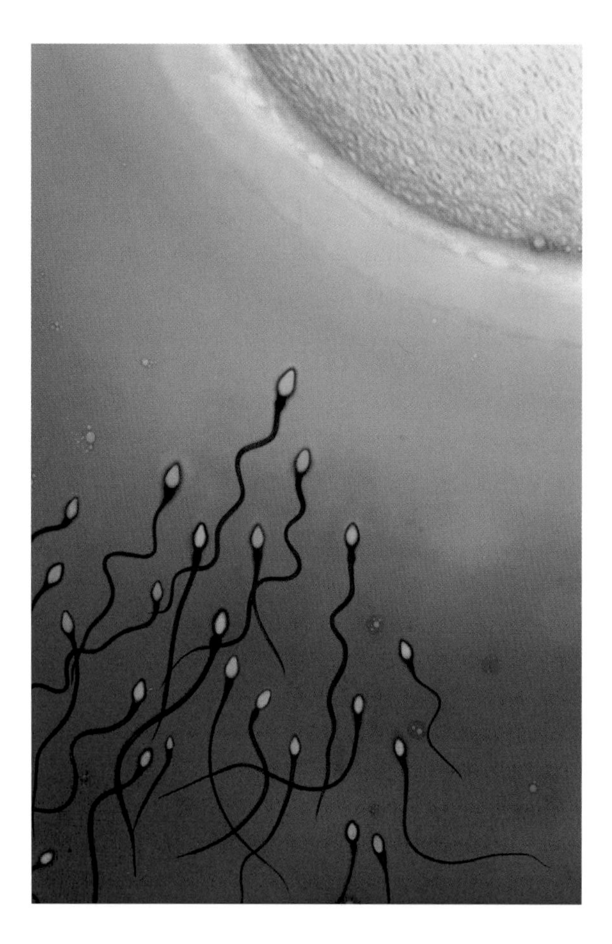

Sperm swim toward an egg. Once a single sperm burrows into the egg, the egg will produce a chemical that prevents more sperm from entering.

colorful, and so on. The mate must choose one suitor over any other. Any individuals that do not make the grade will not mate and, as a result, their genes will not be passed on. That individual may continue to live successfully, but its genes will have been eliminated from future generations.

Once paired up, both individuals have to be able to produce sperm or eggs that are capable of working properly. Fertilization has to take place and the developing eggs have to be protected. Fertilization may take place internally (inside the female's body) or externally (outside the body). Internal fertilization provides more protection for the sperm and eggs because they are not exposed to the elements.

And, finally, the young have to survive in order for the parents' genes to survive. Any number of things could disrupt the process at any time. Natural selection favors the genes that provide success in all of these areas.

DO NOT GO NEAR THE WATER

The best way to appreciate the diversity of reproduction strategies in living things is to focus on the many modes used by just one group of animals, such as frogs. Many people are familiar with how some frogs reproduce. They lay their eggs in ponds, and the eggs hatch into free-swimming **larvae** called tadpoles or pollywogs. Eventually, the tadpoles go through **metamorphosis** to become adult frogs and crawl out of the water. But there are thousands of different kinds of frogs in the world, and many of them do not lay their eggs in ponds. In fact, frogs reproduce in more different ways than any other kind of mammal, bird, reptile, or fish. Ponds are dangerous places. Eggs and tadpoles are heavily preyed upon by aquatic insects, fish, snakes, and even other tadpoles. Competition with other aquatic species for space and food is high.

As a result, natural selection has favored gene changes in many frog species that take the eggs, tadpoles, or both away from the pond and onto land. As a result, many frog species have become less aquatic and more terrestrial over time. In some cases, the free-swimming larval stage—the tadpole or pollywog—has been eliminated all together.

VIDEO CLIP

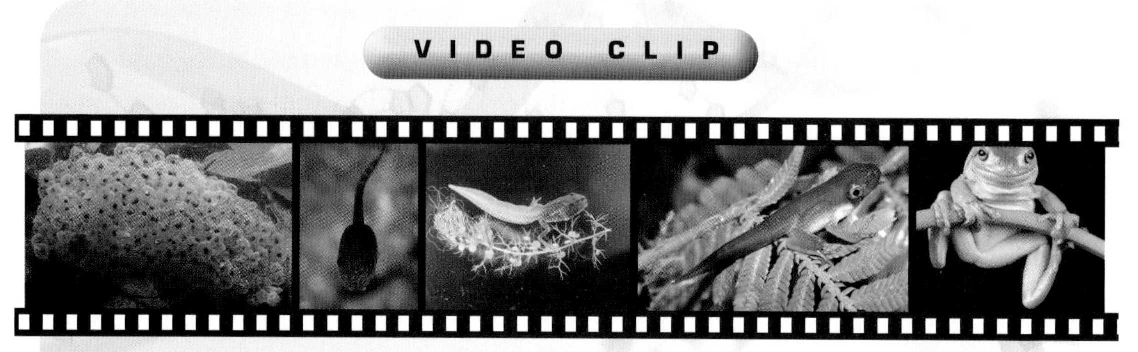

Depending on the species, five stages of frog reproduction and metamorphosis might include *(from left to right)*: 1. eggs that are laid in water, 2. tadpoles with internal gills that live in water, 3. tadpoles that grow legs, 4. froglets that leave the water as their tail disappears and their lungs develop, and 5. adult frogs.

This trend has allowed frogs to spread out into a wider variety of habitats where they can avoid competition with aquatic frogs. They can also avoid a certain level of predation. The basic genetic structure of frogs has

A blue-footed booby performs a courtship dance.

remained the same—a frog still looks like a frog—but the way they reproduce ranges from the ordinary to the bizarre.

Reproduction always begins with attracting a mate. Male birds are often brightly colored, as attractive to the dull-colored females as a Christmas tree is to a small child. Male birds sing. Monkeys hoot and howl. Horses whinny, neigh, and prance. Many animals dance, and some, such as penguins, give gifts of food.

And frogs call. Depending on the species, they croak, whistle, click, ring, bellow, squeak, hum, trill, or boom like a big bass drum. Each species has its own specific call so that females are only attracted to males of their own species. A frog's call is a strong factor in natural selection, because any females attracted to the wrong male will not get their eggs fertilized and, therefore, will not pass on their genes.

The first step toward a more terrestrial life for many frog species has been to deposit their eggs out of the water. They lay many eggs with small yolks because the eggs only need enough energy to hatch into tadpoles. The tadpoles then return to the water by various means to finish their development into frogs. The tadpoles still have to face predation in the pond, but more eggs survive, which increases the survival rate of the species.

Many species, such as white-lipped frogs from Central America and many true frogs (from a large family of frogs called

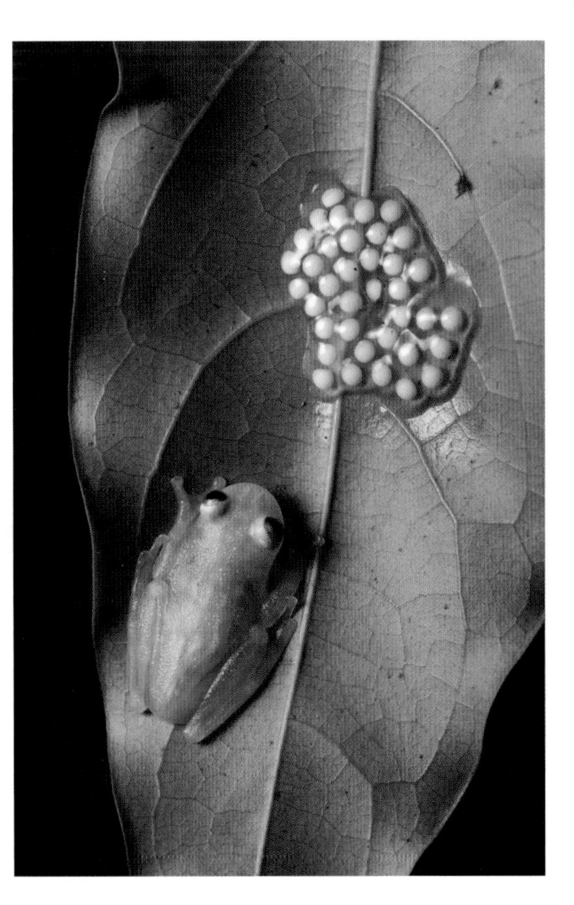

Many species of frogs, such as the glass frog, lay their eggs out of the water on leaves.

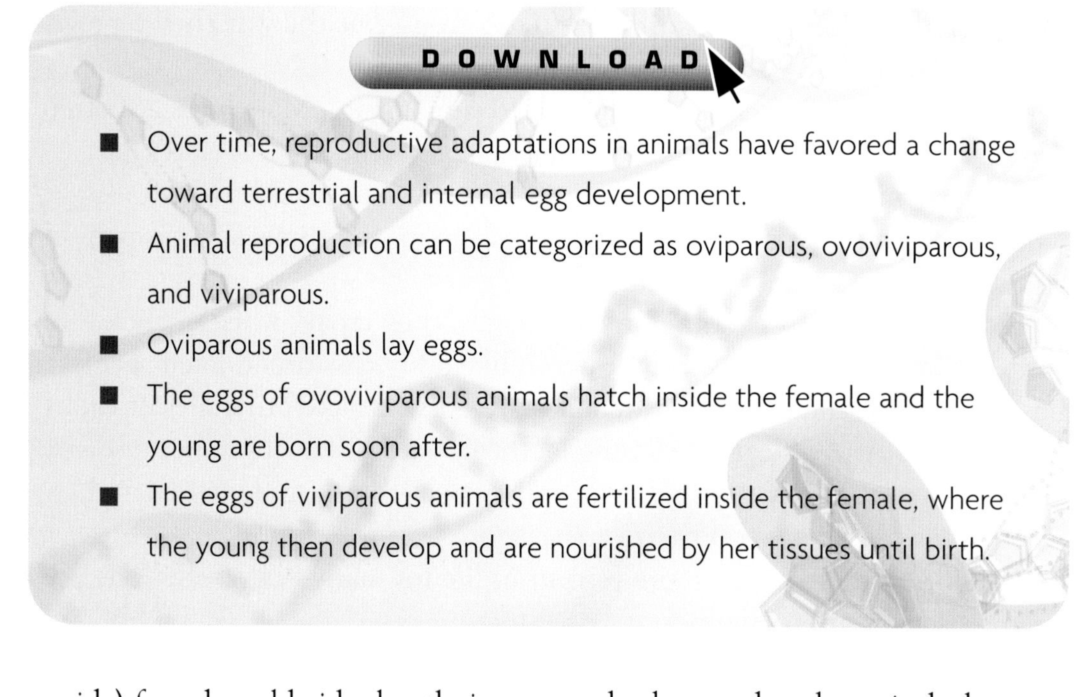

DOWNLOAD

- Over time, reproductive adaptations in animals have favored a change toward terrestrial and internal egg development.
- Animal reproduction can be categorized as oviparous, ovoviviparous, and viviparous.
- Oviparous animals lay eggs.
- The eggs of ovoviviparous animals hatch inside the female and the young are born soon after.
- The eggs of viviparous animals are fertilized inside the female, where the young then develop and are nourished by her tissues until birth.

ranids) found worldwide, lay their eggs under logs and rocks or in holes on the ground. Some lay their eggs on the ground, kicking leaves and dirt over them for camouflage. Some scoop out their own miniature pools in the mud. Once the eggs hatch, the tadpoles are carried to ponds by heavy rains. The glass frog of Mexico and some other tree frogs lay their eggs on leaves, branches, or vines above the water. When the tiny larvae hatch, they fall like raindrops into the water below.

The flying frogs of Africa and Asia (they do not really fly, they are just great leapers, and flaps of skin between their toes allow them to glide a bit) take it a step further. The male grabs the female from behind and clings to her back in typical frog fashion, but then the female climbs onto a branch overhanging the water. As she lays her eggs, the male kicks his legs back and forth, beating up a foamy mound of egg whites like an egg-beater. Inside this gooey nest are hundreds of eggs. The foam, like the meringue on top of a lemon pie, stays moist on the inside and dries to a crisp crust on the outside.

The nest becomes a humidity chamber for the development of the eggs. When the larvae hatch, they excrete a chemical that dissolves the nest, and they drop into the water. Other species beat up a foam nest that floats on top of the water like a boat, effectively removing the eggs from aquatic predators. When the tadpoles hatch, they simply swim down into the water.

Flowering plants reproduce by producing seeds. But the young of plants cannot get up and walk away the way an animal can. As a result, plants have developed a number of successful adaptations that allow them to spread their seeds. The spreading of seeds is called dispersal. Some seeds grow wing-like structures or silky extensions that act like parachutes to carry them away on the wind. Other seeds develop spikes or tiny hooks that get stuck to an animal's fur. The seeds are then carried away by the animal. Some seeds develop pods that pop or explode, shooting the seeds away. Many seeds are sealed inside a colorful, tasty fruit. When an animal eats the fruit, it carries away the seeds, which are dropped again when the animal defecates, or gets rid of waste from the bowels. In fact, some seeds will not even germinate and grow unless the outer coat of the seed is first eroded away by digestion in an animal's stomach.

Dandelion seeds are carried away by the wind *(left)*. Milkweed seeds burst from a pod *(right)*.

All animal species have adaptations to protect their eggs from predation. Birds lay their eggs in nests at the tops of trees, lizards and snakes bury their eggs under the ground, and most mammals carry their eggs inside of their body. The next adaptive step goes beyond egg care and extends to the care of the young.

PROTECTIVE PARENTS

In many animal species, the young are able to survive without the help of their parents. As soon as they are hatched or born, they go out on their own. Any gene changes that favor an increased amount of the time with a protective parent, however, would increase a youngster's chance of survival. Those genes would then be more likely to be passed on.

Parental care may extend from a few weeks to many years. The young are often born or hatched in an early state of development that would not allow them to survive on their own. They are small and defenseless. Chicks are hatched without feathers. Human babies cannot walk or talk. And frog larvae, although they can swim in a pond, have no arms or legs for terrestrial life. Any tadpoles adapted for life on land have to be cared for by a parent.

Frog caretakers fall into two groups: those that care for the eggs and those that extend their care to the tadpole stage. A number of species deposit their eggs in damp places on land—under leaves, underground, in cups of water in rocks and logs, in the centers of cuplike plants—and then stand guard over them until they hatch.

The shovel-nosed frog of Africa, a pudgy frog with a hard, triangularly shaped snout, uses its head like a shovel to burrow into the bank of a pool. Inside this miniature cave, the eggs are laid. The mother stays with them until they hatch. She then tunnels through the earth and creates a slide into the pool. Behind her, the tadpoles wriggle and squirm along the slide until they reach the water.

The midwife toad of Europe does not even leave its eggs. The male fertilizes the eggs, which are strung together like a beaded necklace, as they leave

A pileated woodpecker feeds its chicks in a nest cavity. Birds provide extended care for their young, an adaptation that increases their chicks' chances of survival.

the female's body. Then the male sticks his hind legs into the egg mass and wriggles the eggs up around his waist. He carries the eggs with him wherever he goes, wearing his children like a belt. When the eggs are ready to hatch, he sits patiently in a quiet pond until the tadpoles swim away.

The Surinam toads of South America are flat, bizarre-looking frogs that are fully adapted for aquatic life. Their skin is camouflaged to look like the muddy waters in which they live. Their fingers are long with tiny, touch-sensitive tentacles at their tips for probing for food in the mud. Although the frog rarely leaves the water, it has developed a unique adaptation to protect its eggs. It carries them piggyback!

The male comes up behind the female and grabs her around the waist. As the eggs are laid and fertilized by the male, he uses his hind legs to help scoop them up onto the female's back. The eggs then sink into the soft, spongy skin of her back, where they are surrounded like chicken eggs in an egg carton. The eggs stay in her skin until they are ready to hatch. The tadpoles then pop out of her back one by one and swim away. In one species of Surinam toad, the babies emerge as fully formed frogs. In this way, both the eggs and tadpoles are protected.

The marsupial frogs of Central and South America are the kangaroos of the frog world. Females have a brood pouch on their back, formed from folds in the skin. (The kangaroo's pouch is, of course, on its belly.) The opening of the frog's pouch is close to a chamber called the **cloaca**, from which the eggs are laid. As the male fertilizes the eggs, he shovels them into the pouch with his feet.

As the eggs develop, the female's back bulges out, looking as if a cluster of grapes has been slipped under her skin. When the eggs hatch, she sits in a pond. Like the rubber man in the circus, she reaches up with her back legs and pulls open the pouch with her toes. Out swim the tadpoles or, in some species, fully developed froglets.

By extending care through the tadpole stage, many frogs gain an even greater adaptive advantage, ensuring that more young survive to the next generation. These frogs usually lay fewer eggs, because fewer will be lost to predation. Often the yolks are larger because the developing tadpoles are not getting as much of the energy that they need from feeding in ponds.

The poison arrow (or poison dart) frogs found in the tropical forests of Central and South America have become totally terrestrial, living on or close to the ground. They never enter ponds to breed. Brightly colored in red, orange, yellow, fluorescent green, and blue, depending on the species, they advertise their poisonous skins. These frogs lay only a few eggs on leaves or on the ground. The male guards the eggs, visiting pools of water to suck water up into his cloaca. Returning to his eggs, he squirts them with the water to keep them moist.

When the eggs hatch, the male (and sometimes the female) sticks his legs into the egg mass and a couple of tadpoles wriggle up onto his back. The skin on the frog's back excretes a sticky glue to help the tadpoles hang on. Then the parent carries its young to a cup of water in a plant or rock. In some species, the female visits the cup from time to time to feed the tadpoles by laying special **infertile** eggs in the water. The tadpoles then eat the infertile eggs.

GIVING BIRTH

Animals produce young in one of three ways: **oviparous** *(oh-VIP-uh-ruhs)*, producing eggs that develop and hatch outside of the mother's body; **ovoviviparous** *(oh-voh-vy-VIP-uh-rhus)*, producing eggs that develop within the mother's body and hatch within or immediately after being laid outside the body; and **viviparous** *(vy-VIP-uh-rhus)*, producing eggs that develop entirely inside the female. The developing young of viviparous animals are nourished by the female's tissues until they are born "live."

Nearly all mammals, many reptiles, and some fishes are viviparous, but only three species of closely related toads are. Female African live-bearing toads hold their eggs inside their body for nine months—the same amount of time as humans do—until fully formed miniature toads are born. Natural selection has eliminated both the egg-laying and tadpole stages.

However, live birth has been accomplished by some frogs that are actually oviparous. Females of the mouth-brooding frog of South America lay their eggs on the ground, where they are fertilized and guarded for up to three weeks by more than one male. When the developing larvae

The garter snake is ovoviviparous. It produces eggs that develop without a shell inside the female's body. When the young are "born," they break out of thin egg sacs *(upper left)*. The common eider is oviparous. Its eggs develop and hatch outside of the female's body *(lower left)*. Horses are viviparous. Their young develop inside of the female's body and are born live.

inside the eggs begin to wiggle, each male leans down and pops a few into his mouth. But instead of being swallowed, the eggs slide into the vocal sac in the parent's throat. The vocal sac is the voice chamber that frogs use to attract their mates. The tadpoles will hatch inside the vocal sacs and live there until fully developed, bean-sized froglets hop out of the fathers' mouths.

The female Australian gastric brooding frog takes it a step further. At breeding time, the female stops eating and genes signal the digestive juices that normally enter her stomach to stop flowing. After laying her

Unlike most mammals, the platypus and echidnas, or spiny anteaters, of Australia and New Guinea are oviparous—they lay eggs like a bird instead of giving birth to live young. Although egg laying is a primitive characteristic that makes them more closely related to extinct mammals, these animals have other characteristics that are highly developed. The platypus has a duck-like bill that it uses to dig for prey, such as worms and freshwater shrimp, along the dark, mucky bottoms of streams and rivers. The bill contains special sensors that can detect electrical currents given off by a prey's muscle movements when the prey tries to escape. The echidnas have a narrow snout and a long, sticky tongue specialized for eating ants and termites. Like other mammals, both the platypus and the echidna have hair and produce milk to feed their young.

eggs, she eats them. In her stomach, the eggs remain undigested, where they hatch and grow into froglets. Then she belches up her young, and the tiny frogs pop from her mouth like miniature jack-in-the-boxes. After the young are "born," the production of digestive juices resumes, and the female begins eating again.

Changes in reproductive strategies in both plants and animals have allowed them to take advantage of a great variety of habitats, which reduces competition and fosters species diversity. Listing every adaptation in every living species would fill volumes. And that does not even take into account the number of defensive or feeding strategies, or any other kinds of adaptations. Just think of the human with its estimated 25,000 genes. That is 25,000 specific adaptations for just one species. And scientists estimate that there are at least 5 million and possibly as many as 30 million species on Earth. Well, you get the picture.

HUMAN DEVELOPMENT AND PARENTAL CARE IN HUMANS

Life is hazardous, and for a species to survive, the individuals of that species must produce enough young to replace natural losses in the population. This generally happens in one of two ways. One strategy is to produce lots of young with little or no care so that at least some of the young will survive natural hazards, such as predation. Another is to produce few young with extended parental care so that fewer young are lost.

Many fish are a good example of the first strategy, laying hundreds, thousands, or even millions of eggs at a time outside of their bodies and then abandoning them. When the young hatch, they know instinctively how to find food and take care of themselves.

Mammals, including humans, are a good example of the second strategy. They retain their young inside of their bodies for part of the young's development. Inside the mother's body, the young are protected from environmental hazards and from predators. When the young are born, they are often hairless and blind. They cannot care for themselves. The parents continue to care for them, protecting them, feeding them, and keeping them warm long after birth.

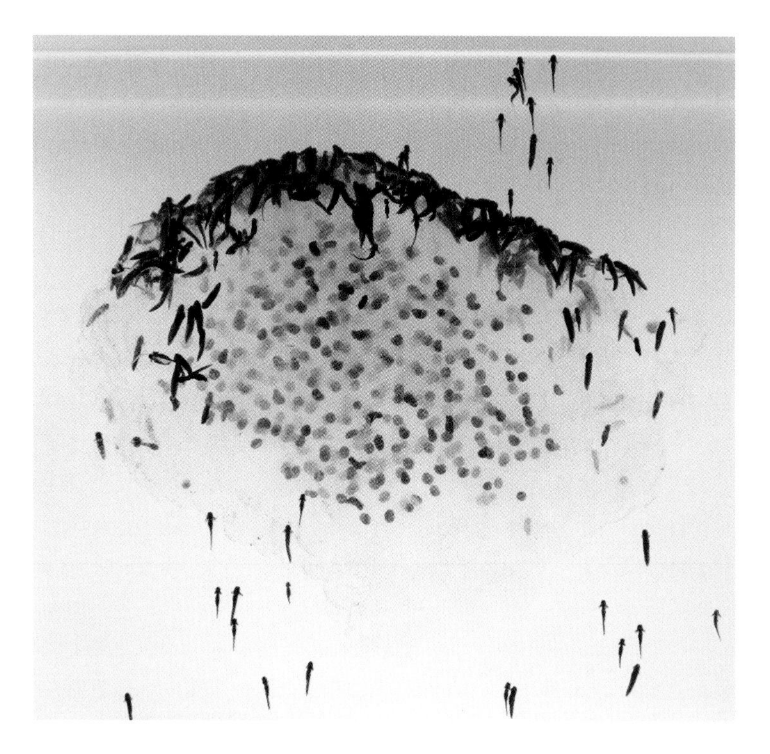

Many frogs produce large numbers of young, like these tadpoles hatching from an egg mass, which are not cared for by the parents.

As the young grow and develop, they become less and less dependent upon their parents, although in some species that may take a while. Humans, for example, provide parental care for at least 18, if not more, years!

While parental care benefits a species by ensuring that a greater number of young will survive, it comes at a cost to the parents. The time and energy required to raise young is much greater than if the young are left to fend on their own. On the other hand, a species that does not provide parental care uses more energy to produce a greater number of young, even though they do not expend any additional energy in raising the young afterward.

Humans produce few young, usually only one at a time, which are cared for over many years.

Glossary

adaptation—a trait that makes a living thing better able to survive

artificial selection—the process whereby human beings choose individual plants or animals with desired traits to serve as parents for the next generation; selective breeding

asexual reproduction—reproduction involving only one parent in which an identical set of the parent's DNA is passed on to the offspring

binomial nomenclature—the naming system in which each organism is given a two-part name, reflecting the genus and species

biodiversity—variation that is found in living organisms, their genes, and their habitats

bioluminescence—light given off from a living organism

carnivorous—feeding on animals

cells—the basic units of life

classify—to arrange organisms into groups based on their similarities

cloaca—the common chamber in amphibians, reptiles, birds, and many fish into which body wastes, sperm, and eggs empty before leaving the body

commensalism—interaction between two different species in which one species benefits while the other is neither helped nor harmed

cryptic coloration—coloration that allows an organism to blend into its background or environment, thus making it less likely it will be preyed upon

disruptive coloration—colors and patterns that disrupt or break up the shape or outline of an organism's body

domesticated—tamed or cultivated for use by human beings

ectoparasite—a parasite that lives on the outside of a host's body

elliptical—shaped like an oval instead of a round circle

endoparasite—a parasite that lives on the inside of a host's body

fertilization—the union of a sperm and an egg to produce a new individual

flash coloration—vivid coloration that is revealed suddenly by an organism under attack only after its cryptic or camouflage coloration fails

gene flow—the physical movement of genes in or out of a population as individuals either leave or enter

genetic diversity—a type of biodiversity that describes the different kinds of genes found in an individual or within a population of the same species

genetic drift—a random change in the frequency of genes in a population because of chance events

genus—a scientific group that includes a number of similar, closely related species

gestalt—something that is recognized as a whole by the sum of all its parts, so that if one part is changed, the whole is no longer recognizable

gradualism—when species change slowly and continuously over a long period of time

hybrid infertility—the inability of individuals created by crossing two different species to produce young themselves

infertile—not able to produce offspring

innate—not learned; controlled by reflex or instinct

isolating mechanism—any barrier (environmental, behavioral, or reproductive) that keeps genes from being exchanged between individuals in different populations

larvae—the early form of any insect or animal that changes into another form when it becomes an adult

meiosis—cell division that produces gametes or sex cells, each with half the chromosomes of the original cell; also called reduction division

metamorphosis—an organism's change in form from an early stage to an adult stage, such as the change of a tadpole into a frog

mimicry—the way in which one organism looks like another organism or like some natural object

mutualism—interaction between two different species in which both species benefit

natural selection—the process by which individuals that are better adapted to their environment are more likely to survive, reproduce, and pass on their genes than others of their same kind

organism—a living thing

oviparous—producing eggs that develop and hatch outside of the female's body

ovoviviparous—producing eggs that develop inside of the female's body and then hatch immediately before or soon after laying

parasitism—interaction between two different species in which one species benefits while the other is harmed

photosynthesis—the production of sugar in plants using energy from the sun

physiology—the science that deals with the way in which the parts of living things work

predator—an animal that lives by catching and eating other animals

prey—an animal that is hunted for food by another animal

punctuated equilibrium—long periods of species stability interrupted by short periods of rapid change during which new species form

respire—to take up oxygen, burn sugar, and release carbon dioxide; includes breathing in animals with lungs and the exchange of oxygen and carbon dioxide in cells (cellular respiration)

self-pollinate—to transfer pollen from the male part of one flower to the female part of the same flower, or to a flower on the same plant

sensory—of the senses (sight, hearing, touch, taste, and smell)

sexual reproduction—reproduction involving two parents in which each parent contributes half his or her DNA to produce a new and unique individual

speciation—the process by which new species of organisms are formed

species—a group of organisms that can mate with each other and produce fertile offspring

species diversity—the number and variety of species found in a certain environment or particular location

symbiosis—the living together of two different species in which one or both have become totally dependent on the other for survival

terrestrial—living on land

vestigial—a part of the body that is small and not as fully developed as it once was in the ancestors of a species

viviparous—giving birth to live young instead of producing eggs that hatch

Search Engine

BOOKS

Cochran, Doris M. *Living Amphibians of the World*. Garden City, NY: Doubleday, 1961.

Duellman, William E., and Linda Trueb. *Biology of Amphibians*. New York: McGraw-Hill, 1986.

Owen, Jennifer. *Feeding Strategy*. Chicago: University of Chicago Press, 1980.

Schafer, Susan. *Snakes* (Perfect Pets series). Tarrytown, NY: Marshall Cavendish, 2003.

WEB SITES

University of California at Berkeley
http://evolution.berkeley.edu/evolibrary/article//evo_31

Encyclopedia of Earth
www.eoearth.org/search?q=biodiversity

Grain Valley School District, Animals and Animal Adaptations
http://grainvalley.k12.mo.us/gvsd/eagle_resources/science.html

How Stuff Works "Science Channel" (Search keywords "animal adaptations" and "parasites")
http://science.howstuffworks.com

Makalapa Elementary School, Adaptations and Survival
www.makalapa.k12.hi.us/Makalapa_Folder/HTML/adapt&survive/adapt&survive.html

Index

Page numbers in italics refer to illustrations.

About the Author

Susan Schafer is a science teacher and the author of several nonfiction books for children. She has written about numerous animals, including horses, snakes, tigers, Komodo dragons, and Galapagos tortoises. Her book on the latter was named an Outstanding Science Trade Book for Children by the National Science Teachers Association and Children's Book Council. She has also written a fictional book about animal tails for very young children. Schafer has spent many years working in the field of biology and enjoys sharing her knowledge and appreciation of nature with others. She lives on a ranch in Santa Margarita, California, with her husband, horses, and dogs, and with the beauty of the oak-covered hills around her.